Library of
Davidson College

Titles in This Series

10 **Joel Friedman, Editor,** Expanding Graphs
9 **William T. Trotter, Editor,** Planar Graphs
8 **Simon Gindikin, Editor,** Mathematical Methods of Analysis of Biopolymer Sequences
7 **Lyle A. McGeoch and Daniel D. Sleator, Editors,** On-Line Algorithms
6 **Jacob E. Goodman, Richard Pollack, and William Steiger, Editors,** Discrete and Computational Geometry: Papers from the DIMACS Special Year
5 **Frank Hwang, Clyde Monma, and Fred Roberts, Editors,** Reliability of Computer and Communication Networks
4 **Peter Gritzmann and Bernd Sturmfels, Editors,** Applied Geometry and Discrete Mathematics, The Victor Klee Festschrift
3 **E. M. Clarke and R. P. Kurshan, Editors,** Computer-Aided Verification '90
2 **Joan Feigenbaum and Michael Merritt, Editors,** Distributed Computing and Cryptography
1 **William Cook and Paul D. Seymour, Editors,** Polyhedral Combinatorics

DIMACS
Series in Discrete Mathematics
and Theoretical Computer Science

Volume 10

Expanding Graphs

Proceedings of a DIMACS Workshop
May 11–14, 1992

Joel Friedman
Editor

NSF Science and Technology Center
in Discrete Mathematics and Theoretical Computer Science
A consortium of Rutgers University, Princeton University,
AT&T Bell Labs, Bellcore

American Mathematical Society

This DIMACS volume resulting from the Special Year on Graph Theory and Algorithms contains research articles and extended abstracts from participants at the Expander Graphs Workshop held at DIMACS from May 11, 1992, through May 14, 1992.

1991 *Mathematics Subject Classification.* Primary 05–06, 05C35, 05C80, 05C85, 60Jxx, 68Rxx.

Library of Congress Cataloging-in-Publication Data

Expanding graphs/Joel Friedman, editor.
    p. cm.—(DIMACS series in discrete mathematics and theoretical computer science, ISSN 1052-1798; v. 10)
    Papers from the DIMACS Workshop on Expander Graphs, May 11–14, 1992, Princeton University.
    Includes bibliographical references.
    ISBN 0-8218-6602-8
    1. Graph theory—Congresses. I. Friedman, Joel. II. Series.
QA166.E97   1993   93-4708
511′.5–dc20   CIP

Copying and reprinting. Individual readers of this publication, and nonprofit libraries acting for them, are permitted to make fair use of the material, such as to copy an article for use in teaching or research. Permission is granted to quote brief passages from this publication in reviews, provided the customary acknowledgment of the source is given.

Republication, systematic copying, or multiple reproduction of any material in this publication (including abstracts) is permitted only under license from the American Mathematical Society. Requests for such permission should be addressed to the Manager of Editorial Services, American Mathematical Society, P.O. Box 6248, Providence, Rhode Island 02940-6248.

The appearance of the code on the first page of an article in this book indicates the copyright owner's consent for copying beyond that permitted by Sections 107 or 108 of the U.S. Copyright Law, provided that the fee of $1.00 plus $.25 per page for each copy be paid directly to the Copyright Clearance Center, Inc., 27 Congress Street, Salem, Massachusetts 01970. This consent does not extend to other kinds of copying, such as copying for general distribution, for advertising or promotional purposes, for creating new collective works, or for resale.

        Copyright ©1993 by the American Mathematical Society. All rights reserved.
        The American Mathematical Society retains all rights except those granted
                                        to the United States Government.
                                  Printed in the United States of America.
        The paper used in this book is acid-free and falls within the guidelines
                          established to ensure permanence and durability. ∞

        All articles in this volume were printed from copy prepared by the authors.
            Some articles were typeset using $\mathcal{A}_{\mathcal{M}}\mathcal{S}$-TeX and $\mathcal{A}_{\mathcal{M}}\mathcal{S}$-LaTeX,
                the American Mathematical Society's TeX macro system.

                      10 9 8 7 6 5 4 3 2 1   98 97 96 95 94 93

Contents

Foreword	vii
Preface	ix
Random Cayley Graphs and Expanders (Abstract) NOGA ALON AND YUVAL ROICHMAN	1
Spectral Geometry and the Cheeger Constant ROBERT BROOKS	5
The Laplacian of a Hypergraph FAN R.K. CHUNG	21
Uniform Sampling Modulo a Group of Symmetries Using Markov Chain Simulation MARK JERRUM	37
On the Second Eigenvalue and Linear Expansion of Regular Graphs NABIL KAHALE	49
Numerical Investigation of the Spectrum for Certain Families of Cayley Graphs JOHN LAFFERTY AND DANIEL ROCKMORE	63
Some Algebraic Constructions of Dense Graphs of Large Girth and of Large Size FELIX LAZEBNIK AND VASILIY A. USTIMENKO	75
Groups and Expanders A. LUBOTZKY AND B. WEISS	95
Ramanujan Graphs and Diagrams Function Field Approach MOSHE MORGENSTERN	111
Highly Expanding Graphs Obtained from Dihedral Groups HOLGER SCHELLWAT	117
Are Finite Upper Half Plane Graphs Ramanujan? AUDREY TERRAS	125

Foreword

This DIMACS volume on expander graphs contains abstracts or papers from talks at a workshop held at DIMACS, May 11–14, 1992.

We would especially like to thank Joel Friedman for putting together this Proceedings that has papers from so many outstanding researchers within this field.

This workshop was part of the DIMACS 1991–1992 Special Year on Graph Theory and Algorithms organized by Fan R. K. Chung and William T. Trotter.

Fred Roberts, Director
Robert Tarjan, Co-Director
Diane Souvaine, Associate Director

Preface

The DIMACS Workshop on Expander Graphs took place at Princeton University, May 11–14, 1992. There were 70 participants. The program featured 22 talks and two open problem sessions. This volume contains much of the material covered at this workshop, in the form of unrefereed papers or summaries of the talks.

The field of expanding graphs involves a number of different fields of study, and gives rise to important connections between them. We were happy to have many of these fields represented at the workshop, including theoretical computer science, combinatorics, probability theory, representation theory, number theory, and differential geometry; the workshop was a wonderful opportunity to assemble researchers and topics with a diversity not usually found in more regular conferences and meetings. We received many positive responses from the participants of the workshop.

We would like to thank the DIMACS executive committee for sponsoring the workshop. Fan Chung, Daniel Gorenstein, Fred Roberts, Bob Tarjan, Tom Trotter, Pat Toci, Carol Rusnak, Adam Buchsbaum, and Ramesh Sitaraman all helped us greatly. We regret the untimely passing of Daniel Gorenstein, whose energies greatly contributed to DIMACS in many ways. Winnie Waring was of especial help in organizing the workshop. Persi Diaconis helped with the proposal for the workshop. We also wish to thank Christine Thivierge and Donna Harmon at the AMS for helping to prepare this volume.

Joel Friedman, Princeton
February 1993

Random Cayley Graphs and Expanders
(Abstract)

NOGA ALON AND YUVAL ROICHMAN

ABSTRACT. For every $1 > \delta > 0$ there exists a $c = c(\delta) > 0$ such that for every group G of order n, and for a set S of $c(\delta) \log n$ random elements in the group, the expected value of the second largest eigenvalue of the normalized adjacency matrix of the Cayley graph $X(G, S)$ is at most $(1 - \delta)$. This implies that almost every such a graph is an $\varepsilon(\delta)$-expander. For Abelian groups this is essentially tight, and explicit constructions can be given in some cases.

The (undirected) Cayley Graph $X(G, S)$ of a group G with respect to the set S of elements in the group is the graph whose set of vertices is G and whose set of edges is the set of all (unordered) pairs $\{\{g, gs\} | g \in G, s \in S\}$. Obviously, this is a regular graph of degree at most $2|S|$ and hence its diameter is at least $log_{2|S|}|G|$. Babai and Erdös [4] proved that every group of order n has $\log_2 n + O(\log \log n)$ elements $x_1, \ldots, x_t$ such that every group element is a product of the form $x_1^{\varepsilon_1} \ldots x_t^{\varepsilon_t}, \varepsilon_i \in \{0, 1\}$. It follows that G has a set of $log_2 n + O(\log \log n)$ generators such that the resulting Cayley graph has a logarithmic diameter. In [1] it is proved that the diameter is almost surely logarithmic with respect to $c \log n$ random elements, for an appropriate absolute constant c. For more details see [4] and [1]. For a general survey on Cayley graphs with small diameters see [5].

Definition. A graph H is called a c-expander if for every set of vertices S,

$$|\Gamma(S)| > c|S|(1 - |S|/|H|)),$$

where $\Gamma(S)$ is the set of all neighbours of S.

Department of Mathematics, Raymond and Beverly Sackler Faculty of Exact Sciences, Tel Aviv University, Ramat Aviv, Tel Aviv, Israel. Research supported in part by a U.S.A.-Israeli BSF grant.

Department of Mathematics, Hebrew University of Jerusalem, Jerusalem, Israel. 1991 Mathematics Subject Classification. Primary 05C25, 05C80. The detailed version of this paper has been submitted for publication elsewhere.

It is obvious that any c-expander (for some fixed $c > 0$) has a logarithmic diameter, but the converse does not hold, as shown, for example, by the Cayley graph of the symmetric group with respect to the set of all transpositions. For a graph H, let $\mu_1[H]$ denote the second largest eigenvalue in absolute value of the adjacency matrix A_H of H. If H is d-regular, the normalized adjacency matrix A_H^* of H is the doubly stochastic matrix $\frac{1}{d}A_H$. Let $\mu_1^*[H]$ denote the second largest eigenvalue in absolute value of A_H^*. Clearly $\mu_1^*[H] = \frac{1}{d}\mu_1[H]$. Our first result is the following:

THEOREM 1. *For every $1 > \delta > 0$ there is a $c(\delta) > 0$ such that the following holds. Let G be a group of order n and let S be a set of $c(\delta)\log_2 n$ random elements of G. Then*
$$E\{|\mu_1^*[X(G,S)]|\} < 1 - \delta.$$

By considering an appropriate martingale as done in [6] and by the easy direction of the well known connection between the second eigenvalue of a graph and its expansion properties (see [2] and [7]), this implies the following.

COROLLARY 2. *For every $1 > \varepsilon > 0$ there exists a $c(\varepsilon) > 0$ such that the following holds. Let G be a group of order n, and let S be a random set of $c(\varepsilon)\log_2 n$ elements of G, then the Cayley graph $X(G,S)$ is an ε-expander almost surely. (I.e., the probability it is such an expander tends to 1 as n tends to infinity.)*

The $\log n$ term in the last corollary is needed in the following strong sense for every Abelian group:

PROPOSITION 3. *For every fixed $\delta > 0$ there is a constant $c = c(\delta) > 0$ such that the following holds. If A is an Abelian group of order n, and $X(A,S)$ is a δ-expander, then $|S| \geq c\log_2 n$.*

For Abelian groups the estimate in Theorem 1 can be improved as follows.

THEOREM 4. *For each δ, $0 < \delta < 1$, there exists an $\varepsilon = O(\delta\log(1/\delta))$ such that for any Abelian group A of order n the Cayley graph $X(A,S)$ of A with respect to a subset S of $(1+\varepsilon)\log_2 n$ (not necessarily distinct) random elements satisfies $|\mu_1^*[X(A,S)]| \leq 1 - \delta$ almost surely.*

This result is tight, as shown by the next statement.

PROPOSITION 5. *Suppose $G = Z_2^m$ and let $X(G,S)$ be a Cayley graph of G whose second largest eigenvalue in absolute value is at most $(1-\delta)|S|$. Then*
$$|S| \geq (1 + \Omega(\delta\log(\frac{1}{\delta})))m.$$

We also discuss some explicit constructions for the groups Z_2^m by applying techniques from the theory of error correcting codes in order to estimate the relevant eigenvalues. A representative example is the following result, proved by applying the constructions of [3]

PROPOSITION 6. *There exists an absolute constant $c > 0$ such that for every $\epsilon > 0$ and for every m one can describe explicitly a Cayley graph $H = X(Z_2^m, S_m)$ where $l = |S_m| \leq cm^2/\epsilon^2$, so that $|\mu_1[H]| \leq \epsilon |S_m| = O(\sqrt{l}m)$.*

The proofs, together with various related results, will appear in the full version of the paper.

References

1. N. Alon, A. Barak and U. Manber, On disseminating information reliably without broadcasting, Proc. of the 7th International Conference on Distributed Computing Systems (ICDS), Berlin, September 1987, pp. 74-81.
2. N. Alon and V. D. Milman, λ_1, isoperimetric inequalities for graphs and superconcentrators, J. Comb. Theory, Ser.B, 38 (1985), 73–88. See also: N. Alon and V. D. Milman, Eigenvalues, expanders and superconcentrators, Proc. 25th IEEE FOCS, Singer Island, Florida, IEEE(1984), 320-322.
3. N. Alon, O. Goldreich, J. Hastad and R. Peralta, Simple constructions of almost k-wise independent random variables, Proc. 31st IEEE FOCS, St. Louis, Missouri, IEEE (1990), 544-553. See also: Random Structures and Algorithms 3 (1992), 289-304.
4. L. Babai and P. Erdös, Representation of group elements as short products, in: Theory and Practice of Combinatorics, G. Sabidussi, editor, Ann. Disc. Math. 12 (1982), 27–30.
5. L. Babai, G. Hetyei, W. M. Kantor, A. Lubotzky and A. Seress, On the diameter of finite groups, Proc. 31st IEEE FOCS, IEEE (1990), 857-865.
6. A. Broder and E. Shamir, On the second eigenvalue of random regular graphs, Proc. 28th IEEE FOCS, IEEE (1987), 286-294.
7. R. M. Tanner, Explicit construction of concentrators from generalized N-gons, SIAM J. Alg. Disc. Meth. 5 (1984), 287-293.

Spectral Geometry and the Cheeger Constant

ROBERT BROOKS

July, 1992

Perhaps the most basic inequality in spectral geometry is the following, due to Jeff Cheeger:

Theorem [Ch]: $\lambda_1 \geq (1/4)h^2$.

Here, λ_1 is the first eigenvalue of the Laplacian

$$\Delta(f) = -\mathrm{div}(\mathrm{grad}(f))$$

on a closed Riemannian manifold M, $\lambda_0 = 0$ being the zero-th eigenvalue, and h is the Cheeger constant

$$h(M) = \inf_S \frac{\mathrm{area}(S)}{\min\left[\mathrm{vol}(A),\mathrm{vol}(B)\right]},$$

where S runs over hypersurfaces dividing M into two parts A and B.

If M is an n-dimensional manifold, "area" denotes $(n-1)$-dimensional volume, while "vol" denotes n-dimensional volume. Of course, when $n=2$, $(n-1)$-dimensional volume is usually called "length", while n-dimensional volume is usually called "area" rather than "volume," but that should cause little confusion.

Cheeger's inequality is not difficult to prove, and indeed there are versions of Cheeger's inequality in the non-compact case and in the case of manifolds with boundary, with either Dirichlet or Neumann boundary conditions. Of particular interest to us here is the non-compact case

$$\lambda_0 \geq (1/4)h^2,$$

1991 *Mathematics Subject Classification.* 58G99.
Partially supported by NSF grants DMS-9000631 and DMS-9200313
The final version of this paper will be submitted for publication elsewhere

where
$$h(M) = \inf_S \frac{\text{area}(S)}{\text{vol}(\text{int}(S))},$$
where S runs over hypersurfaces cutting M into a compact piece $\text{int}(S)$ and an unbounded piece, and where λ_0 is the bottom of the L^2 spectrum on M.

Cheeger's inequality has an important converse, due to Peter Buser [Bu], which states:

Theorem ([Bu]): There exist constants c_1 and c_2, depending on a lower bound on the Ricci curvature of M, such that
$$\lambda_1 \leq c_1 \cdot h + c_2 \cdot h^2.$$

The constants produced by Buser's proof are not particularly good, but Buser's inequality says that, in the presence of bounded curvature, λ_1 and h are qualitatively the same thing.

It is not surprising that there are graph-theoretic analogues of these notions and results. As usual, there are different conventions and normalizations which are essentially equivalent. We find it convenient to stick to the case of k-regular graphs Γ, and define the Laplacian as
$$\Delta(f)(x) = \frac{1}{k} \sum_{y \sim x} [f(x) - f(y)],$$
and the Cheeger constant to be
$$h(\Gamma) = \inf_E \frac{\#(E)}{\min(\#(A), \#(B))},$$
where E where E runs over collections of edges such that $\Gamma - E$ decomposes into two pieces A and B, and $\#(A)$ (resp. $\#(B)$) denotes the number of vertices in A (resp. B).

Cheeger's inequality then becomes:

Theorem: $\lambda_1 \geq \frac{1}{2k^2} h^2$.

The analogue of Buser's inequality is:

Theorem: $\lambda_1 \leq 2h$.

One of our interests in the past few years (see, e.g., [CPSG], [SGTC], [kReg], and [SGN]), has been to pass back and forth between the geometric and graph-theoretic pictures. One reason for doing this is that a problem which apears difficult from one point of view may be relatively easy, or even solved, from the other point of view. Another reason is that attitudes towards various results may differ markedly in the two areas, and comparing them may be an important source of insight.

The present paper arises from comparing differing points of view on Cheeger's inequality. The geometric point of view is to regard Cheeger's inequality as an estimate for λ_1 in terms of h. While it may be difficult in particular instances to get a real qualitative hold of h (see [BW] for an exception to this), the attitude generally is that h is fairly easy to understand from a qualitative point of view, and gives a reasonably strong link between analysis and geometry.

The attitude in computer science, however, seems to be quite different, see [Bi] for a discussion. There, h tends to be the main object of interest, while λ_1 can often be accessible by other means, for instance via number-theoretic constructions. For example, it is difficult to see how to construct expander graphs, that is, graphs with large h, without first passing through Ramanujan graphs, which are graphs with large λ_1.

It is therefore an interesting question whether spectral methods can be pushed farther to give more information about the Cheeger constant. The results which we give here will be in the direction of a negative answer to this question. That is, we will show that there are intrinsic obstacles to bounding the Cheeger constant in terms of the spectrum, although we will be more qualitative than quantitative about this. We will also give our results in the setting of geometry, although our techniques will be fairly graph-theoretic.

More precisely, we will discuss the following two theorems, proved in [BPP]:

THEOREM 0.1. *There exist two isospectral Riemann surfaces S_1 and S_2, such that S_1 is isospectral to S_2, but $h(S_1) \neq h(S_2)$.*

THEOREM 0.2. *There exist Riemann surfaces S_p of arbitrarily large genus, with*
$$\lambda_1(S_p) \geq 3/16,$$
but
$$h(S_p) \leq \frac{3\log(3)}{2\pi}\left(\frac{p-1}{p+1}\right) \sim .52455.$$

This paper could be thought of as kind of a gloss on (a part of) [BPP], but our attempt here is to make this line of thought accessible to graph-theorists and computer scientists.

For some insight into the relationship between the kinds of questions raised here and computer science, we heartily recommend the paper [Bi]. Indeed, Theorems 0.1 and 0.2 answer questions which were originally raised in [Bi].

Acknowledgements: We would like to thank Joel Friedman for his kind invitation to participate in the DIMACS workshop on "Expanding Graphs and Applications," and to the computer science community for the interest they have shown in these questions.

1. Graph Theory and Geometry

In this section, we discuss the dictionary between graphs and manifolds. The basics of this dictionary are fairly well understood.

Suppose M is a fixed compact manifold. If M' is a covering of M, then M' is described by giving a subgroup $\pi_1(M') \subset \pi_1(M)$. To obtain M' from M, we could proceed in the following way: Let us cut up M by dividing it along hypersurfaces, until we obtain a simply-connected manifold F. Then $\pi_1(M)$ is generated by paths which move precisely once across the cuts we made in M. Let us call these generators $g_1, \ldots, g_k$.

We may then form the Cayley graph of $\pi_1(M)/\pi_1(M')$ relative to the generators $g_1, \ldots, g_k$, whose vertices are the cosets of $\pi_1(M)/\pi_1(M')$, and where two vertices are joined by an edge if the corresponding elements of $\pi_1(M)/\pi_1(M')$ differ by left-multiplication by one of the g_i's.

To reconstruct M' from this graph, we take one copy of F for each vertex, and glue copies of F together according to the following recipe: for each edge, corresponding to a generator g_i, we glue the corresponding copies of F together along the cut that g_i crosses.

We have illustrated this procedure for a simple covering of a torus in Figures 1 and 2 below.

How much of the structure of M' is visible in the graph Γ? Clearly, nothing that has to do directly with F survives restricting to Γ, but perhaps other properties do. For instance, such properties as the volume and diameter of M' are reflected in the volume (i.e. number of vertices) and diameter of Γ, after multiplying by a suitable constant reflecting the geometry of F. Our main observation here is that the same is true for λ_1:

Theorem[SGTC]: For $\lambda_1(M')$ not too large, there exist constants c_1 and c_2, depending only on F, such that

$$c_1 \lambda_1(\Gamma) \leq \lambda_1(M') \leq c_2 \lambda_1(\Gamma).$$

One inequality is fairly easy to see. Given an eigenfunction ϕ of Γ, one can construct a test function f_ϕ on M' which is essentially constant on copies of F, and which changes linearly only near the cuts, in order to join two copies of F which have different constants. The Rayleigh quotient of f_ϕ will then approximate the Rayleigh quotient of f, up to a constant depending only on F.

This gives the inequality $\lambda_1(M') \leq c_2 \lambda_1(\Gamma)$.

There are at least two routes to the opposite inequality. One passes through the Cheeger constant, and is the approach of [SGTC]. Basically, one shows via some geometric measure theory that, up to geometric constants which depend only on F, the most efficient way of dividing M' into two pieces can be approximated by dividing M' up into two sets of copies of F, which then is the same thing as dividing Γ into two pieces.

Another route to this result is contained in the thesis of Marc Burger ([Bur] and [Bur2]). He considers the map from functions on M' to functions on Γ given

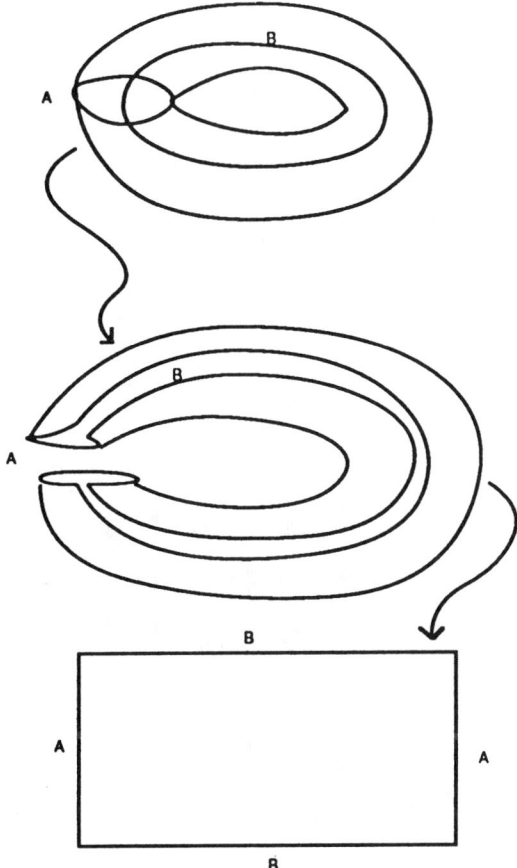

FIGURE 1. Cutting open a torus

by

$$G(x) = \int_{x \cdot F} g \, dx,$$

where $x \cdot F$ is the copy of F corresponding to x.

Roughly speaking, a function g on M' with large Rayleigh quotient can give rise to a function G on Γ which has small Rayleigh quotient, or even which is 0, but only if the average of g over copies of F is small. If g is an eigenfunction, this will imply that the eigenvalue of g must be large.

Let us now compute some particular cases.

Taking first the case of Euclidean n-space $\mathbb{R}^n$, one can readily see that $h(\mathbb{R}^n) = 0$. This is true because a sphere of radius r has area $(\text{const})r^{n-1}$, while it encloses a volume of $(\text{const})r^n$, so that the ratio is $(\text{const})1/r$, which goes to 0 as $r \to \infty$.

It is also not hard to see that $\lambda_0(\mathbb{R}^n) = 0$. To see this, observe that the

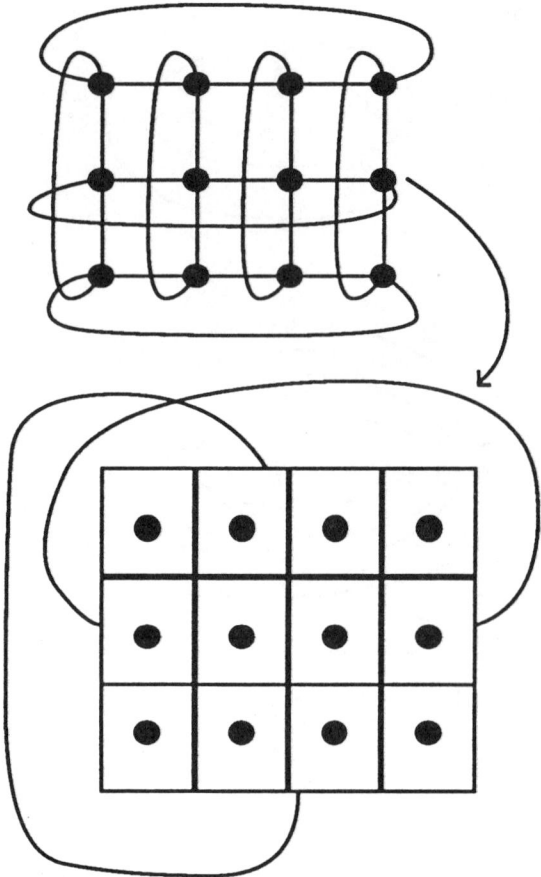

FIGURE 2. Assembling the covering

function $\cos(x/m)$ satisfies

$$-d^2/dx^2(\cos(x/m)) = (\frac{1}{m^2})\cos(x/m),$$

so that

$$g(x) = \Pi_i \cos(x_i/m)$$

has eigenvalue $\frac{n}{m^2}$, which tends to 0 as $m \to \infty$. Restricting g to the interior of the box $\{x: -\pi \cdot m \leq x_i \leq \pi \cdot m\}$ on which g is zero, we get the desired result.

If we consider now the hyperbolic plane $\mathbb{H}^2$, we see that the situation changes. Some standard hyperbolic trigonometry shows that the length of a circle of radius r grows like e^r, and the same is true for the volume it encloses. That tells us that $h(\mathbb{H}^2)$ is at most 1.

To see that this is indeed sharp, one could use a standard symmetrization argument to see that circles are really the best candidates for h. Hence $h(\mathbb{H}^2) = 1$.

Calculating $\lambda_0(\mathbb{H}^2)$ is a little trickier. For one thing, the product construction we made use of in Euclidean space is no longer available. However, we can use the following device, which in retrospect was also available in Euclidean space:

Suppose $f(x)$ is any function which satisfies $\Delta(f) = \lambda \cdot f$, and suppose that x_0 is a point at which $f(x_0) \neq 0$. We may then average f over circles concentric about x_0 to get a function f^{av} which depends only on the distance to x_0, and which is again an eigenfunction with eigenvalue λ.

It follows that f^{av} is a solution to a differential equation in r, and satisfies the initial condition $f'(0) = 0$. In the case of $\mathbb{H}^2$, the differential equation is the equation

$$f'' + \coth(r) f' + \lambda f = 0.$$

It is not difficult to see that λ_0 is the unique value of λ such that this equation has solutions which are positive for $\lambda < \lambda_0$, and has only solutions which take different signs for $\lambda > \lambda_0$. Noting that $\coth(r)$ tends (rather rapidly) to 1, and using a comparison theorem, we calculate that $\lambda_0(\mathbb{H}^2) = 1/4$, since the equation $X^2 + X + \lambda = 0$ has real roots only for $\lambda < 1/4$. This shows that Cheeger's inequality is sharp for H^2.

The solutions $S_\lambda(r)$ of these equations, normalized by $S_\lambda(0) = 1$, are the spherical functions of $\mathbb{H}^2$.

The graph-theoretic analogues of the Euclidean and hyperbolic spaces are the k-regular trees T_k, which are the unique (infinite) k-regular graphs with no closed loops. The 2-tree T_2 is just the real line, with a vertex at each integer, and so is the analogue of Euclidean space, and has $h(T_2) = \lambda_0(T_2) = 0$.

For $k \geq 3$, the situation is a little more complicated. First of all, the computation of the isoperimetric constant of balls gives an estimate $h(T_k) \leq (k-2)$, and a simple symmetrization argument shows that this is sharp. Thus, $h(T_k) = k-2$.

On the other hand, the functions S_λ can be calculated explicitly, see [kReg], and one sees that

$$\lambda_0(T_k) = 1 - 2\frac{\sqrt{k-1}}{k}.$$

Roughly speaking, the argument in the hyperbolic case goes over verbatim, with the exception that the differential equation goes over to a difference equation, which is however of constant coefficients, and so may be solved explicitly.

The interesting point here is that Cheeger's inequality is no longer sharp for the trees T_k.

Peter Doyle [Do] has given a sharper version of Cheeger's inequality for the graph-theoretic case, which does indeed give equality for the trees T_k.

See [kReg] for applications of the spherical functions S_λ to the spectral geometry of graphs.

2. Isospectral Graphs and Surfaces

In this section, we will prove Theorem 0.1.

It has been well-known that isospectral graphs exist in abundance for many years. See [Sch] for a historical survey as of 1973. It appears that the search for isospectral graphs was somewhat ad hoc.

The analogous problem in number theory was, however, rather systematically studied, and the following general recipe was quite standard. It was observed by Sunada in [Su] that it gives a quite general approach to solving isospectral problems of various types.

Theorem([Su]): Let G be a group, and H_1 and H_2 two subgroups of G with the following property:

(†) for all $g \varepsilon G, \#([g] \cap H_1) = \#([g] \cap H_2)$.

Then, for any generators $g_1, \ldots, g_k$ for G, the Cayley graphs of G/H_1 and G/H_2 are isospectral.

The condition (†) can be rephrased in a lot of ways. One way is to see that it is the same as saying that $L^2(G/H_1)$ is unitarily equivalent as a G-module to G/H_2.

The G-equivariance is then easily seen to imply that the Laplacian on G/H_1 goes over to the Laplacian of G/H_2, showing that the spectra agree.

It was Sunada's breathtaking observation in [Su] that the same is true for manifolds:

Theorem[Su]: Suppose that (G, H_1, H_2) satisfies (†), and suppose that there is a surjective map $\phi : \pi_1(M) \to G$.

Then the coverings M^{H_1} and M^{H_2} corresponding to $\phi^{-1}(H_1)$ and $\phi^{-1}(H_2)$ respectively are isospectral.

The simplest proof of this to date seems to be via the L^2 characterization of (†), and is due to Buser [Bu]:

We first observe that the Cayley graphs of $\Gamma_1 = \pi_1(M)/\phi^{-1}(H_1)$ and $\Gamma_2 = \pi_1(M)/\phi^{-1}(H_2)$ relative to the generators $g_1, \ldots, g_k$ are the same as the Cayley graphs of G/H_1 and G/H_2 respectively, relative to the generators $\phi(g_1), \ldots, \phi(g_k)$. By (†), L^2 of these Cayley graphs are unitarily equivalent as G-modules.

For each $x \varepsilon \Gamma_1$, let δ_x denote the delta-function on x, and $\psi_x = \psi(\delta_x)$ be the function on Γ_2 given by the unitary equivalence.

Then any function f on M^{H_1} can be written as

$$f = \sum_x f_x \cdot \delta_x,$$

where $f_x = f$ restricted to the copy of F corresponding to x.

The function

$$\psi(f) = \sum_x f_x \cdot \psi_{\delta_x}$$

is then a function on M^{H_2}. G-equivariance shows that $\psi(f)$ is smooth if f is, because moving across a boundary of f corresponds to left-multiplication by a generator, and unitary equivalence implies that Rayleigh quotients are preserved, and that the map ψ is invertible. It follows at once that the spectrum of Δ is preserved.

The condition (†) is simple enough to allow for fairly explicit examples. My personal favorite (see [BT] for why) is the following:

Let G be the group $PSL(3, \mathbb{Z}/2)$, and

$$H_1 = \begin{pmatrix} * & * & * \\ 0 & * & * \\ 0 & * & * \end{pmatrix}, \quad H_2 = \begin{pmatrix} * & 0 & 0 \\ * & * & * \\ * & * & * \end{pmatrix}.$$

Then (G, H_1, H_2) satisfies (†).

A particularly nice choice of generators is:

$$A = \begin{pmatrix} 0 & 1 & 1 \\ 0 & 1 & 0 \\ 1 & 0 & 0 \end{pmatrix}, \quad B = \begin{pmatrix} 1 & 0 & 0 \\ 0 & 0 & 1 \\ 0 & 1 & 1 \end{pmatrix}.$$

The Cayley graphs for G/H_1 and G/H_2 are shown in Figures 3 and 4 below.

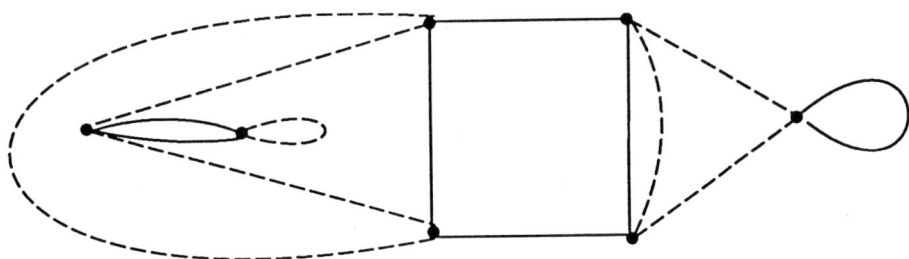

FIGURE 3. The graph Γ_1

It is easily seen that these graphs differ in the Cheeger-like quality that Γ_2 falls into two sizeable pieces by removing one vertex, while Γ_1 does not.

From this, it is an easy matter to construct isospectral Riemann surfaces which model this. The most direct way is to take as our manifold M a two-holed torus with a thin waist, as in Figure 5.

We remark that, by replacing vertices in Figures 3 and 4 with graphs, we may build more complicated graphs which remain isospectral, but whose Cheeger constants differ more and more. We would like to pose the following:

Problem: Maximize the ratio
$$\frac{h(\Gamma_1)}{h(\Gamma_2)}$$

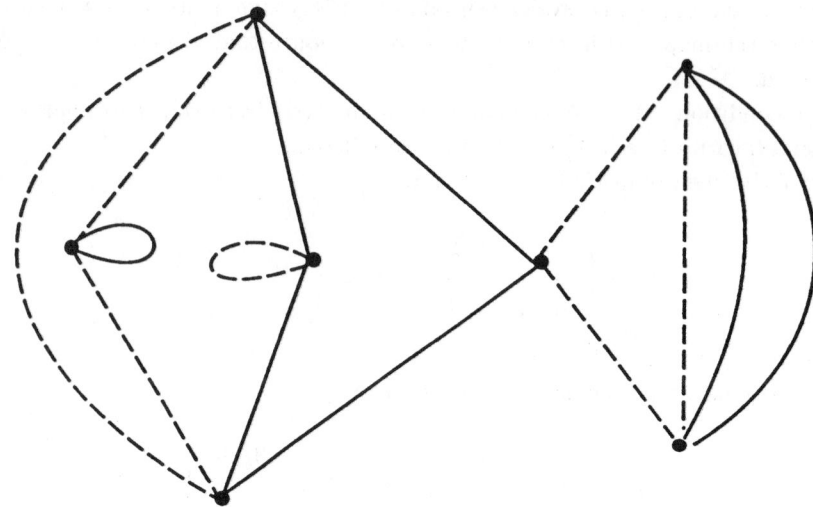

FIGURE 4. The graph Γ_2

for two graphs Γ_1 and Γ_2 which are isospectral.

The inequalities of Cheeger and Buser give an upper bound for this ratio, but we don't know what the sharp upper bound is, or should be.

3. Selberg's 3/16 Theorem

In this section, we prove Theorem 0.2.

We begin with the following theorem, due to Selberg: Consider the group $G = PSL(2, \mathbb{Z})$, and the subgroups

$$G_p = \{\begin{pmatrix} a & b \\ c & d \end{pmatrix} \equiv \pm \begin{pmatrix} 1 & 0 \\ 0 & 1 \end{pmatrix} \pmod{p}\}.$$

We do not assume that p is a prime, but in the discussion below it will be convienient to make that assumption.

Then G acts discretely on $\mathbb{H}^2$, and hence so do the G_p's, so that if $p \neq 2, 3, 4$, or 6, $\mathbb{H}^2/G_p$ is a finite-area Riemann surface S_p with cusps.

We may think of S_p as covering $\mathbb{H}^2/G$ with covering group $PSL(2, \mathbb{Z}/p)$, although strictly speaking this is true only in the "orbifold" sense.

Selberg then showed:

Theorem ([Sel]): $\lambda_1(S_p) \geq 3/16$.

He conjectured that $\lambda_1 \geq 1/4$, this being the best conceivable constant (be-

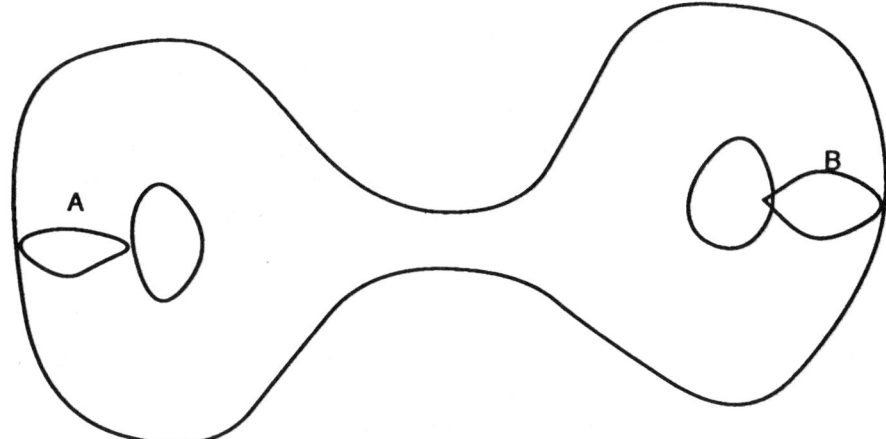

FIGURE 5. The Two-holed Torus

cause $\lambda_0(\mathbb{H}^2) = 1/4$).

These surfaces may be thought of as the first known family of expander graphs, at least after one makes the translations of the dictionary of §1.

To make this more precise, let us pick the generators

$$U = \begin{pmatrix} 1 & 1 \\ 0 & 1 \end{pmatrix} \text{ and } V = \begin{pmatrix} 0 & 1 \\ -1 & 0 \end{pmatrix}$$

for $PSL(2,\mathbb{Z})$. These are the geometric generators for the usual fundamental domain F of $PSL(2,\mathbb{Z})$ in $\mathbb{H}^2$ given by

$$F = \{z : |z| > 1 \text{ and } -1/2 < \operatorname{Re}(z) < 1/2\}.$$

We may then form the graphs Γ_p whose vertices are given by the elements of $PSL(2,\mathbb{Z}/p)$ and whose edges are given by left multiplication by U and V. Since V is of order 2, this is a trivalent graph.

The discussion of §1 then tells us that the graphs Γ_p are expander graphs for some positive constant, this constant essentially being the greatest lower bound of the Cheeger constants of the graphs Γ_p.

Actually, there is a technical point here, which will enter into our discussion in a crucial way. That point comes from the fact that the fundamental domain is non-compact. This problem is discussed and overcome in [DVR].

We have tried for some time to find a direct, elementary proof that the graphs Γ_p are expanders, without first passing through Selberg's Theorem. This would, in turn, give a new, hopefully more conceptual proof of Selberg's Theorem, with, of course, the wrong constant.

This can now be done, see [Hu], [SGN], and especially [Dav].

But for now, let's ask the simpler question: how close do the Cheeger constants $h(S_p)$ come to being 1 (which would give $\lambda_1 \geq 1/4$) or even $\sqrt{3}/2$ (which would give $\lambda_1 \geq 3/16$)?

To that end, let us imagine dividing S_p into two pieces which are unions of copies of F. We do not want to cut along edges which correspond to U, as these are infinitely long. Therefore, we may only cut along the V edges.

We may now model this process on the graphs Γ'_p defined as follows: the vertices of the graph Γ'_p are equivalence classes of ordered pairs (a,b) in $\mathbb{Z}/p \times \mathbb{Z}/p - (0,0)$, with $(a,b) \sim (-a,-b)$, and with (a,b) joined to (a',b') if

$$\det \begin{pmatrix} a & b \\ a' & b' \end{pmatrix} \equiv \pm 1 (\mathrm{mod}\ p).$$

The graphs Γ_p are then p-valent graphs which are then a family of expander graphs, at least in the sense that $\lambda_1(\Gamma'_p)$ is bounded from below. This is pretty much an immediate consequence of Selberg's Theorem, but in this case we have some direct estimates:

Theorem ([BPP]): For p a prime $\equiv 1 (\mathrm{mod}\ 4)$,

$$\frac{p^2 - 2p + 5}{4(p-1)} \leq h(\Gamma'_p) \leq \frac{(p-1)p}{2(p+1)}.$$

Note that as $p \to \infty$, the left-hand side tends to $p/4$, while the right-hand side tends to $p/2$. The p in the numerator comes up because the graphs are p-regular.

Before discussing the proof of this, let us give a way of visualizing these graphs.

We may think of each vertex of Γ'_p as being a cusp of S_p. Indeed, we may consider the hyperbolic surface S'_p built up in the following way: we first construct the regular hyperbolic p-gon H_p whose interior angles are all $2\pi/3$. This will be possible when $p > 6$. We then take one copy of H_p for each vertex in Γ'_p, and glue together two copies of H_p whenever the corresponding vertices are joined by an edge. The surfaces S'_p are then topologically the same as filling in each cusp of S_p by adding a point.

It should certainly be the case that $\lambda_1(S'_p)$ is uniformly bounded from below, but we do not have an honest proof of this.

We remark that the cases $p \leq 6$ are rather classical. When $p < 6$, we may form the spherical regular p-gon with interior angles $2\pi/3$. The resulting surface S'_p is then a sphere, and the graphs Γ'_p are the Platonic solids which are the tetrahedran ($p = 3$), cube($p = 4$), and dodecahedron ($p = 5$). In the case $p = 6$, we let H_p be the regular hexagon in the Euclidean plane, and Γ'_p gives a tessellation pf the torus by hexagons.

Thus, the surfaces S'_p give a natural extension for all p to the series which begins: tetrahedron, cube, dodecahedron, ..., and strikes us as being worthy of study for its own sake.

We also remark that the dual graphs Γ''_p given by the union of the boundaries of copies of H_p in S'_p are trivalent graphs, which must also surely be expander graphs, although we do not have an honest proof of this. This series begins: tetrahedron, octahedron, icosahedron, ...

An important feature of this construction is that the group $PSL(2,\mathbb{Z}/p)$ acts by isometries on S'_p, Γ'_p, and Γ''_p, because it does so on S_p. According to the philosophy of [SGN], this should be helpful in showing that these are expanders.

To prove the lower bound of the Theorem, we observe that the graphs Γ'_p have the following "small diameter" property:

LEMMA 3.1. *Suppose that*

$$\det\begin{pmatrix} a & b \\ a' & b' \end{pmatrix} \neq 0.$$

Then there are precisely two paths of length 2 joining (a,b) and (a',b') in Γ'_p.

The proof is a simple exercise in linear algebra (mod p).

The lower bound now proceed as follows: for each division of the vertices into two sets A and B, with $\#(A) \leq \#(B)$, we count how many paths of length 2 must join an element of A with an element of B. To each such path, one of the two edges must lie in E. We then divide by the number of times each such edge may occur in a path of length 2.

Tp prove the upper bound of the theorem, we need only find an appropriate division of Γ_p into two parts. One choice is:

$$A = \{(a,b) : a \neq 0 \text{ is a square (mod } p)\}$$

$$\cup \{(0,b) : b \text{ is a square (mod } p)\},$$

$$B = \{(a,b) : a \neq 0 \text{ is not a square (mod } p)\}$$

$$\cup \{(0,b) : b \text{ is not a square (mod } p)\}.$$

The condition $p \equiv 1 (\text{mod } 4)$ is required to make this definition well-defined on equivalence classes.

The dividing edge set E is then easily calculated, giving the upper bound.

To conclude the proof of Theorem 0.2, it remains to use the sets A and B above to give a division of S_p into two pieces which are unions of fundamental domains. As the finite arc of the boundary has length $\log(3)$, while F has area $\pi/3$ (these are both elementary calculations in hyperbolic trigonometry), we get the desired upper estimate for $h(S_p)$ given by Theorem 0.2.

REFERENCES

: [Bi] F. Bien, "Construction of Telephone Networks by Group Representations," Notices AMS 36 (1989), pp. 5-22.

: [CIM] R. Brooks, "Constructing Isospectral Manifolds," Am. Math. Month. 95 (1988), pp. 823-839.

- [CPSG] R. Brooks, "Combinatorial Problems in Spectral Geometry," in *Curvature and Topology of Riemannian Manifolds*, Springer Lecture Notes 1201 (1988), pp.14-32.
- [DVR] R. Brooks, "Some Remarks on Volume and Diameter of Riemannian Manifolds," J. Diff. Geom. 27 (1988), pp. 81-86.
- [FGSL] R. Brooks, "The Fundamental Group and the Spectrum of the Laplacian," Comm. Math. Helv. 56 (1981), pp. 581-596.
- [IRLE] R. Brooks, Injectivity Radius and Low Eigenvalues of Hyperbolic Manifolds," J. Reine und Ang. Math. 390 (1988), pp. 117-129.
- [SGN] R. Brooks, "Some Relations Between Spectral Geometry and Number Theory," to appear in Topology 90: Proc. Ohio State University.
- [SGTC] R. Brooks, "The Spectral Geometry of a Tower of Coverings," J. Diff. Geom. 23 (1986), pp. 97-107.
- [kReg] R. Brooks, "The spectral Geometry of k-regular Graphs," to appear in J. d'Analyse.
- [BPP] R. Brooks, P. Perry, and P. Petersen, "On Cheeger's Inequality," to appear.
- [BT] R. Brooks and R. Tse, "Isospectral Surfaces of Small Genus," Nagoya Math. J. 107 (1987), pp. 13-24.
- [BW] R. Brooks and P. Waksman, "The First Eigenvalue of a Scalene Triangle," Proc. AMS 100 (1987), pp. 175-182.
- [Bur] M. Burger, "Petites Valeurs Propres du Laplacien et Topologie de Fell," Thèse, Lausanne, 1986
- [Bur2] M. Burger, "Spectre du Laplacien, Graphes, et Topologie de Fell," Comm. Math. Helv. 63 (1988), pp. 226-252
- [Bu] P. Buser, "A Note on the Isoperimetric Constant," Ann. Sci. Ec. Norm. Sup. 15 (1982), pp. 213-230.
- [Bu 2] P. Buser, "Cayley Graphs and Planar Isospectral Domains," in T. Sunada (ed), *Geometry and Analysis on Manifolds*, Springer Lecture Notes 1339 (1988), pp. 64-77.
- [Bu3] P. Buser, "Isospectral Riemann Surfaces," Ann. Inst. Fourier XXXVI (1986), pp. 167-192.
- [Ch] J. Cheeger, "A Lower Bound for the Smallest Eigenvalue of the Laplacian, " in Gunning (ed), *Problems in Analysis*, Princeton University Press (1970), pp. 195-199.
- [Dav] G. Davidoff, to appear.
- [Do] P. Doyle, personal communication.
- [Hu] M. N. Huxley, "Exceptional eigenvalues and congruence subgroups," in Hejhal, Sarnak, and Terras, *The Selberg Trace Formula and Related Topics*, Contemp. Math 53 (1986), 341-349.
- [Per] R. Perlis, "On the Equation $\zeta_K(s) = \zeta_{K'}(s)$," J. Number Th. 9 (1977), pp. 342-360.

: [Sch] A. Schwenk, "Almost All Trees are Cospectral," in Harary (ed.), *New Directions in Graph Theory,* Academic Press, 1973, pp. 275-307.
: [Se] A. Selberg, "On the Estimation of Fourier Coefficients of Modular Forms," Proc. Symp. Pure Math. VIII (1965), pp. 1-15.
: [Su] T. Sunada, "Riemannian Coverings and Isospectral manifolds," Ann. Math. 121 (1985), pp. 169 - 186.

DEPARTMENT OF MATHEMATICS, UNIVERSITY OF SOUTHERN CALIFORNIA, LOS ANGELES, CALIFORNIA 90089-1113

E-mail address: rbrooksmtha.usc.edu

The Laplacian of a Hypergraph

FAN R. K. CHUNG

March 29, 1993

1. Introduction

Suppose G is a graph with node set N and edge set E consisting of unordered pairs of N. The Laplacian of G, denoted by $L(G)$, is defined to be $D - A$ where A is the adjacency matrix of G (i.e., $A_{ij} = 1$ if $\{i,j\}$ is in E and 0 otherwise), and D is a diagonal matrix with $(D)_{ii} = d(i)$, the degree of the i-th node. Laplacians and the distribution of their eigenvalues imply many important properties of graphs [7,14,15,18,22,29], and lead to many applications in a variety of areas [2,6,13,27,28,32,34,35]. A natural generalization of graphs are so-called hypergraphs. In particular, a k-uniform hypergraph (or, a k-graph for short) has a node set N and edges consisting of k-subsets of N. (Thus, ordinary graphs are 2-graphs.) Many attempts have been made to define the analogue of the Laplacian for hypergraphs and/or some notion of eigenvalues of k-graphs [3,16]. However, various obstructions seem to make the generalization to k-graphs difficult.

In this paper, we will define the Laplacian of a k-graph by considering various homological aspects of hypergraphs. The eigenvalues of the Laplacians will be examined and relations to the other graph properties will be derived. In particular, the eigenvalues of some specified hypergraphs will be evaluated.

In an earlier paper [10], the cohomological aspects of hypergraphs over the finite field Z_2 (and, in general, Z_p) were investigated. The Laplacians for the case of Z_2 have quite different properties from the Laplacians considered here. In this paper, since the operations are over C, the field of complex numbers, the eigenvalues of the Laplacian can be considered. This paper is organized as follows. The definition of the Laplacian will be given in Section 2. The homological

---

1991 *Mathematics Subject Classification.* 05C35.
This paper is in final form.

setting will be described in Sections 3 and 4. Various properties of the eigenvalues of the Laplacian are discussed in Section 5. Some special hypergraphs will be discussed in Section 6. In Section 7 we consider the Laplacian of random graphs. In Section 8, relations of the Laplacian to other graph invariants are discussed and further problems are raised.

2. The definition of the Laplacian

Suppose G is a k-graph with node set N and edge set E which is a subset of $\binom{N}{k}$, the set of all k-sets of N. For a $(k-1)$-subset x of N, its degree, denoted by $d(x)$, is $|\{y \in E : x \subset y\}|$. The average degree d is $\frac{1}{\binom{n}{k-1}} \sum_{x \in \binom{N}{k-1}} d(x)$ where the cardinality of N is n. We say G is d-regular if $d(x) = d$ for all x. The Laplacian of G involves the following matrices whose columns and rows are indexed by $[\substack{N \\ k-1}]$ the set of $(k-1)$-tuples of distinct elements in N:

(i): $D = D(G)$, the diagonal matrix. For $x \in [\substack{N \\ k-1}]$, $D(x,x)$ is defined to be $d(x)$. Also, $D(x,y) = 0$ if $x \neq y$.

(ii): $A = A(G)$, the adjacency matrix. For $x, y \in [\substack{N \\ k-1}]$, $A(x,y)$ is 1 if $x = x_1 x_2 \ldots x_{k-1}$, $y = y_1 x_2 \ldots x_{k-1}$, where $x_i, y_j \in N$, and $y_1, x_1, \ldots, x_{k-1}$ is in E and 0 otherwise. Therefore, $A(x,x)$ is zero.

(iii): $K = K(G)$, the complete graph. That is, for $x, y \in [\substack{N \\ k-1}]$, $K(x,y)$ is 1 if $x = x_1 x_2 \ldots x_{k-1}$ and $y = y_1 x_2 \ldots x_{k-1}$ for $x_1 \neq y_1$ and 0 otherwise.

(iv): $I = I_{k-1}$, the identity matrix.

Now we are ready to define the Laplacian $L(G)$ of a k-graph G, where $k \geq 3$, as follows:

$$L(G) = D - A + \rho(K + (k-1)I).$$

where $\rho = d/n$ is called the density of G.

Sometimes we write $\hat{K} = K + (k-1)I$ and $L(G) = D - A + \rho \hat{K}$ for a k-graph G, $k \geq 3$. When $k = 2$, we have $L(G) = D - A$. As we shall see, the Laplacian corresponds in a natural way to a self-adjoint operator $\partial \delta + \rho \delta \partial$ for some simplicial complex where ρ is a positive constant (uniquely determined by the graph G), as we will discuss in Section 3 and 4. Furthermore, the eigenvalues of the Laplacian defined above play an important role in capturing the essential properties of the graphs.

Lemma 2.1. If $k \geq 3$, $L(G)$ has an eigenvalue ρn with the corresponding eigenvector having all coordinates 1's.

Let f_1 denote the vector with all coordinates 1's. Since we have

$$\begin{aligned} (D - A)f_1 &= 0, \\ (K + (k-1)I)f_1 &= nf_1, \end{aligned}$$

we conclude that $Lf_1 = \rho n f_1$, and f_1 is a eigenvector of $L(G)$ with eigenvalue ρn.

Let $\lambda_1 \geq \lambda_2 \geq \ldots \geq \lambda_{\binom{n}{k-1}}$ denote the eigenvalues of the adjacency matrix A of a d-regular k-graph G. The well-known results of Perron-Frobenius [17,21,31] state that $\lambda_1 = d$ and for $i \neq 1$, $|\lambda_i| \leq d$ (except when A is reducible). Since the eigenvectors associated with $\lambda_i, i \neq 1$, are orthogonal to the all 1's vector, it is easy to see that the Laplacian $L(G)$ has eigenvalues $d, d - \lambda_2, \ldots, d - \lambda_n$.

For a general k-graph G, we define λ_i so that $L(G)$ has eigenvalues $d - \lambda_i$ where d is the average degree. Let us define the spectral value of $L(G)$ to be $\lambda = \lambda(G) = \max_{i \neq 1} |\lambda_i|$. If we can find a good upper bound for λ, then several isoperimetric properties of the k-graphs can be derived. For example, for 2-graphs, the "smallness" of λ implies various properties of the graphs such as: the expansion property (each subset of the node has "many" neighbors), the discrepancy property (each subset induces about the average number of edges), among others. The reader is referred to [7,11] for more details on 2-graphs.

3. The Laplacian of 2-graphs

The Laplacian of a 2-graph is quite simple in comparison to the Laplacian for general k-graphs. Still, it is helpful to review the homological setting for the case of $k = 2$. Throughout this section G is a 2-graph with node set $N = N(G)$ and edge set $E = E(G)$ consisting of unordered pairs of N. We can define the 1-simplicial complex C_1 to be a vector space over R generated by all ordered pairs of N. In addition, we require $(u,v) = -(v,u)$. (Namely, C_1 can be viewed as an exterior algebra with $u \wedge v = -v \wedge u = (u,v)$. We will not use "$\wedge$" notation here.) The boundary operator ∂ is defined by $\partial(u,v) = u - v$ and

$$C_1 \underset{\delta}{\overset{\partial}{\rightleftarrows}} C_0$$

In other words, $\partial : C_1 \to C_0$ can also be interpreted as a matrix W of size $n(n-1) \times n$ where the rows are indexed by $[\binom{N}{2}]$ and the columns are indexed by $[\binom{N}{1}]$. For $x \in [\binom{N}{1}], y \in [\binom{N}{2}]$,

$$W(x,y) = \begin{cases} 1 & \text{if } x = (y,u) \text{ for some } u \\ -1 & \text{if } x = (v,y) \text{ for some } v, \\ 0 & \text{otherwise.} \end{cases}$$

The coboundary operator δ is just the transpose of W.

For a 2-graph G, let $\hat{G}$ denote an orientation of G. That is, each edge (i.e., unordered pair) is assigned a direction (i.e., an ordered pair) and together the directed edges form $E(\hat{G})$. $C_1(\hat{G})$ is defined to be $C_1 \cap E(\hat{G})$ and $C_0(\hat{G}) = \{u \in N : u \text{ is in some edge of } E(G)\}$. As we can see, the Laplacian is independent of the choice of the orientation of G.

The boundary operator ∂ and the coboundary operator δ are just the restriction of the above operators to $C_1(G)$ and $C_0(G)$,

$$C_1(\hat{G}) \underset{\delta}{\overset{\partial}{\rightleftarrows}} C_0(\hat{G}),$$

∂ corresponds to the matrix $W_{(G)}$ with

$$W^{(G)}(x,y) = \begin{cases} W(x,y) & \text{if } x \in C_1(\hat{G}) \text{ and } y \in C_0(\hat{G}) \\ 0 & \text{otherwise} \end{cases}$$

and δ corresponds to the transpose of $W^{(G)}$.

We have $L(G) = \delta\partial = W^{(G)}(W^{(G)})^T = D - A$. Therefore $L(G)$ is semi-positive definite and has one eigenvalue 0 with corresponding eigenvector being the all 1's vector. In general, for a function $f : N \to C$, we have $Lf(v) = \delta\partial(v) = \sum_{u,v \in E(G)} (f(v) - f(u))$. The Laplacians $L(G)$ can be viewed as a discrete analog of the (continuous) Laplace operator on a manifold which maps a function to the corresponding function involving the differences from neighboring points.

4. A homology theory for hypergraphs

Suppose $k > 1$. The simplicial complex C_k is a vector space over R generated by $\begin{bmatrix} N \\ k+1 \end{bmatrix}$ satisfying the property that for f in C_k and $x, y \in \begin{bmatrix} N \\ k+1 \end{bmatrix}$, $f(x) = (-1)^t f(y) = sign(x,y) f(y)$ where t is the number of transpositions in xy^{-1} if x is a permutation of y. For $x \in \begin{bmatrix} N \\ t \end{bmatrix}$ and $y \in \begin{bmatrix} N \\ t-1 \end{bmatrix}$, $sign(x,y)$ is defined to be $sign(x, uy)$ where u is in x but not in y. The boundary operator ∂ can be described as follows:

$$\cdots \underset{\delta}{\overset{\partial}{\rightleftarrows}} C_{k-1} \underset{\delta}{\overset{\partial}{\rightleftarrows}} C_{k-2} \underset{\delta}{\overset{\partial}{\rightleftarrows}} \cdots$$

For x in $C_{k-1}, \partial x = \sum_{y} sign(x,y) y$ where y is a permutation of a $(k-1)$-subset of x.

The coboundary operator δ is the dual of ∂. It is easy to verify the following:
Lemma 4.1. $\partial\partial = 0$ and $\delta\delta = 0$.

Let G be a k-graph. For each edge $x = \{x_1, \ldots, x_k\}$ of G, we can choose a k-tuple $\hat{x} = x_1 \ldots x_k$. These $\hat{x}$ consequently form $C_{k-1}(\hat{G})$ (as it turns out, the choice of the permutations $\hat{x}$ does not affect the Laplacian of G). For $r < k$, $C_r(\hat{G})$ consists of all r-tuples y so that y is contained in ∂x for some x in $C_{r+1}(\hat{G})$.

We consider the Laplacian $\partial\delta + \rho\delta\partial$ restricted to $C_{k-1}(\hat{G}), C_{k-2}(\hat{G})$ and $C_{k-3}(\hat{G})$. It is not difficult to verify the following.
Lemma 4.2 For $y, z \in C_{k-2}(\hat{G})$ $k \geq 3$,

$$\partial \delta z(y) = \begin{cases} \operatorname{sign}(x,y)\operatorname{sign}(x,z) & \text{if the union of the elements of } y \text{ and } z \text{ forms} \\ & \quad \text{an edge } x \text{ of } G, \\ d(y) \cdot \operatorname{sign}(y,z) & \text{if } z \text{ is a permutation of } y, \\ 0 & \text{otherwise}, \end{cases}$$

where $\partial \delta z(y)$ denotes the coefficient of y in $\partial \delta z$. In other words, $\partial \delta z = \sum_y \partial \delta z(y) y$. In a similar way, we have the following:

Lemma 4.3. For $y, z \in C_{k-2}(\hat{G})$,

$$\partial \delta z(y) = \begin{cases} \operatorname{sign}(y,x')\operatorname{sign}(z,x') & \text{if the intersection of elements of } y \text{ and } z \\ & \quad \text{is a } (k-2)\text{-set } x', \\ (k-1)\operatorname{sign}(y,z) & \text{if } z \text{ is a permutation of } y, \\ 0 & \text{otherwise}. \end{cases}$$

Now, let f denote a cochain in $C_{k-2}(G)$. That is, $f = \sum_{x \in \left[\substack{N \\ k-1}\right]} f(x) \cdot x$ and f satisfies $f(x) = \operatorname{sign}(x,y) f(y)$ if y is a permutation of x.

Here we use the following notation: For a $(k-1)$-tuple $x = x_1 \ldots x_{k-1}$ in $\left[\substack{N \\ k-1}\right]$, we define $\bar{x} = \sum_{y, \tilde{y} = \tilde{x}} \operatorname{sign}(y,x) \cdot y$ where $\tilde{x} = x_1, \ldots, x_{k-1}$. Therefore

$$\begin{aligned} f(\bar{x}) &= \sum_{\substack{y \\ \tilde{y} = \tilde{x}}} \operatorname{sign}(y,x) f(y) \\ &= (k-1)! f(x). \end{aligned}$$

We consider the following:

$$\begin{aligned} \partial \delta f(y) &= \sum_{\substack{z \\ zw = \tilde{y}}} d(y) \operatorname{sign}(y,z) f(z) + \sum_{\substack{z \\ \tilde{y} \cup zw = \tilde{x} \in E}} \operatorname{sign}(x,y) \operatorname{sign}(x,z) f(z) \\ &= d(y) f(\bar{y}) + \sum_{\substack{w \in \left[\substack{N \\ k-2}\right] \\ \tilde{w} \subset \tilde{y}}} \sum_{\substack{u \in N - \tilde{y} \\ \tilde{x} = \overline{uy} \in E}} \operatorname{sign}(x,y) \operatorname{sign}(x, uw) f(\overline{uw}) \\ &= Df(\bar{y}) - \sum_{\substack{w \in \left[\substack{N \\ k-2}\right] \\ \tilde{w} \cup v = \tilde{y}}} \operatorname{sign}(y, vw) Af(\overline{vw}) \\ &= Df(\bar{y}) - Af\Big(\sum_{\substack{w \in \left[\substack{N \\ k-2}\right] \\ \tilde{y} = \tilde{w} \cup v}} \operatorname{sign}(y, vw) \overline{vw} \Big) \\ &= Df(\bar{y}) - Af(\bar{y}) \end{aligned}$$

In a similar way, we have

$$
\begin{aligned}
\delta\partial f(y) &= \sum_{\substack{z \\ zw=\bar{y}}} (k-1)\operatorname{sign}(y,z)f(z) + \sum_{\substack{z \\ \bar{y}\cap zw=\bar{w}\in\binom{N}{k-2}}} \operatorname{sign}(y,w)\operatorname{sign}(z,w)f(z) \\
&= (k-1)f(\bar{y}) + \sum_{\substack{w\in[\substack{N\\k-2}] \\ \bar{w}\subset\bar{y}}} \operatorname{sign}(y,w) \sum_{\substack{u\in N-\bar{y} \\ zw=\bar{w}\cup u}} \operatorname{sign}(z,uw)f(z) \\
&= (k-1)f(\bar{y}) + \sum_{\substack{w\in[\substack{N\\k-2}] \\ \bar{w}\subset\bar{y}}} \operatorname{sign}(y,w) \sum_{u\in N-\bar{y}} f(\overline{uw}) \\
&= (k-1)f(\bar{y}) + \sum_{\substack{w\in[\substack{N\\k-2}] \\ \bar{y}=\bar{w}\cup v}} \operatorname{sign}(y,w) K f(\overline{vw}) \\
&= (k-1)f(\bar{y}) + K f(\bar{y})
\end{aligned}
$$

Hence $\delta\partial f(y) = ((k-1)I + K)(\bar{y})$.

Therefore we can view $L(G)$ as $\partial\delta + \rho\delta\partial$ where ρ is the density of G.
In particular, d-regular graphs have edge density $\rho = \frac{d}{n}$ and therefore we have

$$L(G) = dI - A + \frac{d}{n}\hat{K}.$$

We remark that one of the main reasons for the different formulation of the Laplacians for 2-graphs and for k-graphs, $k \geq 3$, is that $C_{k-2}, k \geq 3$ is generated by an oriented basis while the opposite is true for $k = 2$. In fact for $k \geq 3$, the cochains form a vector space of dimension $\binom{n}{k-1}$. Although the Laplacian L has as an eigenvector the all 1's vector (as seen in Lemma 2.1), it is easy to see that all the cochains are orthogonal to the all 1's vector when $k \geq 3$, while this is not so for $k = 2$.

5. Spectral values of the Laplacian

Throughout this section, we again assume G is a d-regular k-graph. We will prove a number of isoperimetric inequalities in terms of the spectral values. First, we prove a lower bound for the spectral value $\lambda(G)$.

THEOREM 5.1. $\lambda(G) \geq \sqrt{d(1-d/n)}$

Proof: We consider $M = dI - L = A - \frac{d}{n}\hat{K}$. It is easy to check that M has eigenvalues $\lambda_i, 1 \leq i \leq [\substack{n\\k-1}]$ where $[\substack{n\\k-1}] = n(n-1)\ldots(n-k-2)$, satisfying $\lambda_1 = 0$ and the spectral value $\lambda = \max_{i\neq 1}\lambda_i$. We consider the trace of MM^T. We have

$$([{}_{k-1}^{n}] - 1)\lambda^2 \geq \sum_i \lambda_i^2$$
$$= TrMM^T$$
$$= d[{}_{k-1}^{n}] - \frac{d^2}{n}[{}_{k-1}^{n}]$$

Therefore
$$\lambda^2 \geq d(1 - d/n)$$

and Theorem 5.1 is proved.

As we will see in Section 7, a random d-regular graph has the spectral value of size $O(\sqrt{d})$. Furthermore several explicit constructions in Section 6 also achieve a spectral value of size $O(\sqrt{d})$, which is within a constant factor of the least possible value.

We now consider using the spectral value to derive isoperimetric inequalities for a graph G.

THEOREM 5.2. *Let S be a subset of the node set N of a d-regular k-graph G. The number $e(S)$ of edges x of G with $x \subseteq S$ satisfies the following:*

$$\left| e(S) - \frac{d}{n}\binom{|S|}{k} \right| \leq \frac{\lambda}{k}\binom{|S|}{k-1} + \frac{d(k-1)}{kn}\binom{|S|}{k-1}$$

Proof: We consider a vector f indexed by $[{}_{k-1}^{N}]$, satisfying $f(x) = 1$ if $\tilde{x} \subseteq S$ and $f(x) = 0$ otherwise. We now consider the bilinear product $\langle f, Lf \rangle$ where $L = L(G)$. It is easily seen that

$$\langle f, Af \rangle = k!e(S).$$

Since $\langle f, \frac{d}{n}\hat{K} f \rangle = \frac{d}{n}k!\binom{|S|}{k} + \frac{d}{n}(k-1)[{}_{k-1}^{|S|}]$, we conclude

$$k!\left(e(S) - \frac{d}{n}\binom{|S|}{k}\right) - \frac{d(k-1)}{n}[{}_{k-1}^{|S|}]$$
$$= \langle f, (L - D)f \rangle$$
$$\leq \lambda\langle f, f \rangle$$
$$= \lambda[{}_{k-1}^{|S|}].$$

Theorem 5.2 is proved.

We note that $\frac{d}{n}\binom{|S|}{k}$ is the expected number of edges contained in $|S|$. The quantity $|e(S) - \frac{d}{n}\binom{|S|}{k}|$ is often called the discrepancy of S, and the discrepancy of G is defined to be the maximum discrepancy over all subsets S, (see [5,7,11] for more discussion on this). The above theorem provides an upper bound for discrepancy in terms of the spectral value.

The inequality in (1) implies that if the spectral value is small in comparison with d, then the number of edges in S is close to the expected quantity. As an immediate consequence, the number of edges involving nodes in S but not entirely contained in S is also close to the expected value (which is almost all the edges involving S, when $|S|$ is small in comparison with n.)

The following isoperimetric equality is slightly stronger than that in Theorem 5.2.

THEOREM 5.3. *Let F denote a subset of $\binom{N}{k-1}$ where N is the node set of a d-regular k-graph G. We define $e_G(F) = |\{(f_1, f_2, x) : f_1, f_2 \in F, f_1 \cup f_2 = x \in E\}|$. Then we have*

$$|e_G(F) - \frac{d}{n}e_K(F)| \leq \frac{\lambda}{k}|F| + \frac{d(k-1)}{kn}|F|$$

where K denotes the complete graph with edge set $\binom{N}{k}$.

Proof: We consider a vector f indexed by $[\binom{N}{k-1}]$ with $f(x) = 1$ if $\tilde{x} \in F$ and 0 otherwise. We consider

$$\langle f, Af \rangle = k! e_G(F).$$

Also,

$$\langle f, \frac{d}{n}\hat{K}f \rangle = \frac{d}{n}k! e_K(F) + \frac{d}{n}(k-1)|F|(k-1)!$$

Therefore

$$k!(e_G(F) - \frac{d}{n}e_K(F) - \frac{d}{kn}(k-1)|F|)$$
$$= \langle f, (L-D)f \rangle$$
$$\leq \lambda \langle f, f \rangle$$
$$= \lambda |F|(k-1)!$$

It would be useful to prove the reverse directions of the above theorem by bounding the spectral value in terms of the discrepancy. However, the following problem for 2-graphs still remains unresolved [7].

Conjecture: Suppose G is a 2-graph satisfying

$$\left| e(S) - \rho\binom{|S|}{2} \right| \leq \alpha |S|$$

where ρ is the density. Is it true that $\lambda(G) \leq c\alpha$ for some absolute constant c?

Now, we consider isoperimetric inequalities of a somewhat different flavor. For a subset S of the node set N of a k-graph G, we define the neighborhood of S as follows:

$\Gamma(S) = \{u \in N : \{u\} \cup w \in E(G) \text{ for some } (k-1)\text{-subset } w \text{ of } S\}$.

THEOREM 5.4. *Let S be a subset of the node set N of a d-regular k-graph G where $k \geq 3$. Then for any $S \subseteq N$, we have*

$$|\Gamma(S)| \geq \frac{d^2 [\genfrac{}{}{0pt}{}{|S|}{k-1}]}{\lambda^2 \left(1 - \frac{[\genfrac{}{}{0pt}{}{|S|}{k-1}]}{[\genfrac{}{}{0pt}{}{n}{k-1}]}\right) + \frac{d^2|S|}{kn}}.$$

Proof: We define a vector f indexed by $[\genfrac{}{}{0pt}{}{N}{k-1}]$, so that $f(x) = 1$ if $\tilde{x} \subseteq S$ and 0 otherwise. Suppose that the eigenvalues of the Laplacian $L(G)$ are $d - \lambda_i$ where $\lambda_1 = 0$ and the orthonormal eigenvectors are denoted by v_i.

Suppose $f = \sum a_i v_i$ and therefore $\sum a_i^2 = \|f\|^2 = [\genfrac{}{}{0pt}{}{|S|}{k-1}]$. We consider the following inner product:

$$\begin{aligned}
\langle\ f(A - \frac{d}{n}\hat{K}), (A - \frac{d}{n}\hat{K})f\rangle & \\
= \langle fA, Af\rangle - 2\langle fA, \frac{d}{n}\hat{K}f\rangle &+ \frac{d^2}{n^2}\langle f\hat{K}, \hat{K}f\rangle \\
= \langle fA, Af\rangle - \frac{d^2}{n^2}\langle f\hat{K}, \hat{K}f\rangle & \\
= \langle fA, Af\rangle - \frac{d^2}{n}[\genfrac{}{}{0pt}{}{|S|}{k}]. &
\end{aligned}$$

On the other hand, we have

$$\begin{aligned}
\langle f(dI - L), (dI - L)f\rangle &= \langle f(A - \frac{d}{n}\hat{K}), (A - \frac{d}{n}\hat{K})f\rangle \\
&= \sum a_i^2 \lambda_i^2 \\
&\leq \lambda^2 \left([\genfrac{}{}{0pt}{}{|S|}{k-1}] - a_1^2\right) \\
&= \lambda^2 \left([\genfrac{}{}{0pt}{}{|S|}{k-1}] - \frac{[\genfrac{}{}{0pt}{}{|S|}{k-1}]^2}{[\genfrac{}{}{0pt}{}{n}{k-1}]}\right)
\end{aligned}$$

Furthermore, we have

$$\begin{aligned}
\langle fA, Af\rangle &= \sum_{w,w' \in [\genfrac{}{}{0pt}{}{S}{k-1}]} |\{u \in N : w = aw'', w' = bw'' \text{ and } \overline{abw''} \in E\}| \\
&= \sum_{u \in N} |\{w \in [\genfrac{}{}{0pt}{}{S}{k-1}] : \overline{uw} \in E\}|^2 \\
&\geq \frac{\left(\sum_{u \in N} |\{w \in [\genfrac{}{}{0pt}{}{S}{k-1}] : \overline{uw} \in E\}|\right)^2}{|\Gamma(S)|} \\
&\geq \frac{\left(\sum_{w \in [\genfrac{}{}{0pt}{}{S}{k-1}]} d(w)\right)^2}{|\Gamma(S)|} \\
&= \frac{d^2 [\genfrac{}{}{0pt}{}{|S|}{k-1}]^2}{|\Gamma(S)|}
\end{aligned}$$

All together, we get

$$|\Gamma(S)| \geq \frac{d^2[\binom{|S|}{k-1}]^2}{\lambda^2\left(\binom{|S|}{k-1} - \frac{[\binom{|S|}{k-1}]^2}{\binom{n}{k-1}}\right) + \frac{d^2}{n}[\binom{|S|}{k}]}$$

$$\geq \frac{d^2[\binom{|S|}{k-1}]}{\lambda^2\left(1 - \frac{[\binom{|S|}{k-1}]}{\binom{n}{k-1}}\right) + \frac{d^2}{n}\frac{|S|}{k}}$$

We note that for the case of $k = 2$, the statement of Theorem 6.6 is also true. It is shown in [1] and [36] that for a d-regular 2-graph G with node set N, for any $S \subseteq N$ we have

$$|\Gamma(S)| \geq \frac{d^2|S|}{\lambda^2\left(1 - \frac{|S|}{n}\right) + d^2\frac{|S|}{n}}$$

where λ is the spectral value of the Laplacian $L(G) = D - A$.

For dense 2-graphs (e.g. having $d > n^{31/32}$) small discrepancy implies certain bounds for the spectral value by using some recent work on "quasi-randomness". The reader is referred to [5,8,11] for the interrelationships of many random-like properties for dense graphs. Here we sketch the connection of spectral value with several invariants related to quasi-randomness.

Let ρ denote the density of a k-graph G. We define a function $f: N \times \ldots \times N \to R$ such that $f(v_1, \ldots, v_k) = 1 - \rho$ if $\{v_1, \ldots, v_k\}$ is an edge of G; and $f(v_1, \ldots, v_k) = -\rho$ otherwise. The 2-deviation of G, denoted by $dev_2(G)$, is defined by

$$(5.1) \quad n^{k+2} dev_2(G) = \sum_{\substack{u_1,v_1,u_2,v_2 \\ u_3,\ldots,u_k \in N}} f(u_1, u_2, v_3, \ldots, v_k) f(u_1, v_2, v_3, \ldots, v_k)$$

$$f(v_1, u_2, v_3, \ldots, v_k) f(v_1, v_2, v_3, \ldots, v_k)$$

THEOREM 5.5.

$$n^{k+2}(dev_2(G) + O(\frac{1}{n})) = \sum_{i=1}^{\binom{n}{k-1}} \lambda_i^4$$

where $d - \lambda$'s are the eigenvalues by the Laplacian.

Proof: This follows from the fact that

$$\sum_i \lambda_i^4 = Tr(dI - L)^4$$

$$= Tr(A - \rho\hat{K})^4$$

$$= n^k + 2(dev_2(G) + O(\frac{1}{n}))$$

since for $x, y \in \binom{N}{k}$, $(L(G))(x,y) = f(u_1, \ldots, u_k)$ where the union of the elements of X and Y is $u_1, \ldots, u_k$.

In [5] it was shown that the discrepancy of G is at least $cn^k(dev_2 G)^4$ and therefore for a 2-graph G, $disc_2 G \geq c\lambda^{16} n^{-14}$ for some constant c. The same approach does not seem to yield nontrivial upper bounds for λ for general $k \geq 3$.

6. Laplacians of Cayley graphs and its generalizations

In this section we consider Laplacians of k-graphs and their generalizations with density ranging from $1/2$ to $n^{-1+1/t}$ for a constant t. As we will see, the spectral value of these explicitly constructed graphs are quite close to the optimum.

First we consider a special type of matrices whose eigenvalues can be easily determined. Let g be a real-valued function defined on Z_p, integers modulo p. The matrix $M_{n,g}^{(k)}$ is of size $n^{k-1} \times n^{k-1}$ with (x,y)-entries to be $g(x_1 + \ldots + x_k)$ for $x = (x_1, \ldots, x_{k-1})$ and $y = (y_1, x_2, \ldots, x_{k-1})$.

THEOREM 6.1. *The matrix $M_{n,g}^{(3)}$ has eigenvalues $\sum_x g(x)$ and $\epsilon |\sum_x g(x)\theta^x|$ where θ is an nth root of unity $\neq 1$ and $\epsilon = 1$ or -1.*

Proof:
Proof: We will first construct vectors using θ and ψ and subsequently show these are indeed eigenvectors of M. We define the vector $f = f_{\theta\psi}$ to be

$$f(x,y) = \theta^x \psi^{y+x} + \theta^{x+y}\psi^x + (\theta^{-x}\psi^{-y-x} + \theta^{-x-y}\psi^{-x}) \cdot \frac{\epsilon \sum_t g(t)(\theta\psi)^t}{|\sum_t g(t)(\theta\psi)^t|}$$

where $\epsilon = 1$ or -1.
Now $(Mf)(y,z) = \sum_x g(x+y+z)f(x,z)$
We note that

$$\sum_x g(x+y+z)\theta^x \psi^{x+z} = \theta^{-y-z}\psi^{-y} \sum_x g(x+y+z)(\theta\psi)^{x+y+z}$$

Therefore

$$Mf_{\theta\psi} = \epsilon |\sum_x g(x)(\theta\psi)^x| f_{\theta\psi}$$

Therefore $f_{\theta\psi}$ are eigenvectors. In fact, $f_{\theta\psi}$ are all the eigenvectors (by considering the rank of M). Theorem 6.1 is proved.

Now we can consider Cayley k-graphs $C_{n,S}$ with node set Z_n and edges $\{x_1, \ldots, x_k\}$ if and only if $x_1 + \ldots + x_k$ is in S for some fixed set S.

THEOREM 6.2. *The matrix C_{ng}^k has eigenvalues $\sum_x g(x)$ and $\epsilon |\sum_x g(x)\theta^x|$ where θ is an nth root of unity $\neq 1$ and $\epsilon = 1$ or -1.*

We note that the above Cayley graph is, in fact, defined with edge set N^k. We remark that hypergraph properties discussed in Section 5 still hold with this slight modification.

THEOREM 6.3. *The Laplacian of the Cayley graph $C_{n,S}^{(k)}$ has spectral value*

$$\max_{\theta \neq 1} |\sum_x \psi_s(x)\theta^x| \text{ where } \psi_S(x) \text{ is } 1-\rho \text{ if } x \in S, \text{ and is } -\rho \text{ if } x \notin S$$

Proof: Let $\tilde\psi_S$ denote the function $\psi_S(x) = 1$ if x is in S and 0 otherwise. Since the Laplacian of $C_{n,S}$ satisfies $L - D = M_{n,\tilde\psi_S} - \frac{|S|}{n} M_{n,\tilde\psi_N}^{(k)}$, using the proof of Theorem 6.1, the eigenvector with the eigenvalue $|S|$ for M_{n,ψ_s}^k also has eigenvalue $|S|$ for $\frac{|S|}{n} M_{n,\tilde\psi_N}^{(k)}$. Therefore $L - D$ has eigenvalues 0 and $\epsilon|\sum_\psi \psi_S(x)\theta^x|$ for $\theta \neq 1$ since $\tilde\psi_S(x) - \frac{\rho}{n}\tilde\psi_N(x) = \psi_S(x)$.

The Paley graph $P_p^{(k)}$ is just a special case of the Cayley graph $C_{n,S}^{(k)}$ taking n to be some prime congruent to 1 mod p and S to be the set of quadratic residues of p. Therefore we have the following:

THEOREM 6.4. *The Laplacian of the Paley graph $P_p^{(k)}$ has spectral value at most $\sqrt{p}/2$.*

Proof: For $\theta \neq 1$, it is well known [23] that

$$|\sum_x \chi(x)\theta^x| \leq \sqrt{p}.$$

where χ is the usual non-principal quadratic character given by $\chi(x) = 1$ if x is a quadratic residue and -1 otherwise. Therefore, $\lambda \leq \sqrt{p}/2$ since $\tilde\psi(x) = \chi(x)/2 + \frac{1}{p}$ where Q denotes the set of quadratic residues.

There are several ways to obtain generalization of Paley k-graphs. Suppose H is a subgroup of $GF(p)^*$ with index α^{-1} (i.e., $\alpha = |H|/(p-1)$). We can construct $P_{p,\alpha}^{(k)}$ to be the Cayley graph $C_{p,H}$ with edge density α.

THEOREM 6.5. *The Laplacian of the generalized Paley graph $P_{p,\alpha}^{(k)}$ has spectral value at most $\sqrt{p}/2$.*

Proof: By Theorem 6.3, we want to bound $\sum_x \psi_S(x)\zeta^{jx}$ for $j \neq 0$ and $\zeta = e^{\frac{2p i i}{p}}$. Let Φ_H denote the set of all nontrivial characters χ from $GF(p)^*$ to C^* such that $\chi|H = 1$ and $\chi(0) = 0$. It is not difficult (see [23,33]) to check that

$$\psi_H = \alpha \sum_{\chi \in \Phi_H} \chi$$

and $|\Phi_H| = \alpha^{-1} - 1$. Therefore we have

$$\left| \sum_{x \in GF(p)} \psi_S(x) \zeta^{jx} \right|$$

$$= \alpha \left| \sum_{x \in GF(p)} \sum_{\chi \in \Phi_H} \chi(x) \zeta^{jx} \right|$$

$$\leq \alpha \sum_{\chi \in \Phi_H} \left| \sum_{x \in GF(p)} \chi(x) \zeta^{jx} \right|$$

$$\leq \alpha |\Phi_H| \sqrt{p}$$

$$\leq (1-\alpha)\sqrt{p}.$$

The proof for Theorem 6.5 is completed.

Another family of constructions are the so-called coset graphs (see [6]). We consider $GF(p^t)$ and let X denote a coset $x + GF(p)$ where $GF(p^t) = GF(p)(x)$ and the coset k-graph $C_{\psi,t}^{(t)}$ is the Cayley graph $C_{p^t,X}^{(k)}$ with edge density $n^{1-\frac{1}{t}}$ where $n = p^t - 1$.

THEOREM 6.6. *The coset k-graph $C_{p,t}^{(k)}$ has $n = p^t - 1$ nodes, degree p and spectral value at most $(t-1)\sqrt{p}$.*

Proof: The proof follows from the following generalization of the character sum inequality which was conjectured in [6] and proved by N. Katz [25]:

$$\left| \sum_{a \in GF(p)} \chi(x+a) \right| \leq (t-1)\sqrt{p}$$

where χ is a nontrivial multiplicative character of $GF(p^t)$.

7. The spectral value of the Laplacian of a random graph

When we say a random graph with density ρ has spectral value $c\sqrt{\rho n}$, we mean that with probability approaching 1 almost all graphs on n nodes with density ρ have spectral value $c\sqrt{\rho n}$ where c denotes some absolute constant. The proofs for determining the spectral value of a random dense graph (ρ being a positive constant) are considerably easier than the proofs for sparse graphs ($\rho = \frac{d}{n}$ for constant degree d).

For sparse graphs, there are basically two different approaches for estimating the spectral value. One usual method is to examine the trace of A^t for some large $t \ll \log n$ (see [4,16,24]) along the line given in Friedman [16]. The second method, used by Kahn and Szemerédi, is to reduce the problem so that the adjacency matrix of the random graph operates on selected finitely many vectors with expected behaviors. Both methods use elaborated techniques and careful analysis. Although similar approaches can be carried on for hypergraphs to obtain an upper bound of $c\sqrt{\rho n}$, we will not give the arguments here. Instead, here we give a short proof of a (weak) upper bound $O(n^{1/2+\epsilon})$ for dense random graphs with constant edge density ρ. We consider A^t for some large constant $t > k$. For $x, y \in \binom{N}{k-1}$, the (x,y)-entry of A^t can be estimated depending on the set intersection of x and y. If x and y share w common elements, the (x,y)-entry of A^t differs from the expected value of $\rho^{t-k+2-w}n^{t-k+2-w}$ by at most $ct\rho^{(t-k+2-w)/2}n^{(t-k+2-w)/2}$. In a similar way, the (x,y)-entry of $\hat{K}^t$ differs from the expected value of $n^{t-k+2-w}$ by at most $c + n^{(t-k+2-w)/2}$. So, let us consider $\frac{\langle f, Mf \rangle}{\langle f, f \rangle}$ for f orthogonal to the all 1's vector where $M = (L-D)^t = (A - \rho\hat{K})^t$. Therefore,

$$\begin{aligned}
\langle f, A^t f \rangle &= \langle f, Mf \rangle. \\
&\leq |\sum_{x,y} M(x,y) f(x) f(y)| \\
&\leq \sum_{x,y} |M(x,y)|(f^2(x) + f^2(y))/2 \\
&\leq ct \left(\sum_{w=0}^{k-1} \binom{n-k+1}{k-1-w} \binom{k-1}{w} \rho^{(t-k+2-w)/2} n^{(t-k+1)/2} \right) \sum_x f^2(x) \\
&\leq c't \left(\rho^{(t-k+2)/2} n^{(t-k+2-w)/2} \right) \langle f, f \rangle.
\end{aligned}$$

Therefore $\lambda^t \leq c't\rho^{(t-k+2)/2} n^{(t-2+w)/2}$. By taking large t, we obtain,

$$\lambda \leq c(\rho n)^{1/2+\epsilon}$$

for any $\epsilon > 0$.

8. Concluding remarks

In previous sections, we have examined relations between the spectral value of the Laplacian and several graph invariants. Numerous questions remains unresolved several of which we mention here.

One natural question is the following: Is it true that small spectral value implies "quasi-randomness" (its definition is given in [5,11,12]). The answer is however negative. It is not difficult to check that the following k-graph with edge density $1/2$ has spectral value at most $csqrtn$ but is not quasi-random. Let $G_{(k-1)}$ denote a random $(k-1)$-graph on n nodes. Construct a k-graph H so that $x = \{x_1, \ldots, x_k\}$ is an edge of H if and only if $|\binom{x}{k-1} \cap E(G_{k-1})|$ is odd. It is shown in [8,12] that H is not quasi-random but it is not difficult to check H has spectral value of the same order as a random k-graph.

One possible definition for a "strong" Laplacian (which can be related to "quasi-randomness") is as follows: Here we will only describe the definition for 3-graphs G while it can be easily generalized to k-graphs. We define $L^\star = D - T_G$ where for $p_1, p_2, p_3 \in \binom{N}{2}$, $T_G(p_1, p_2, p_3)$ is defined to be $1 - \rho$ if the union of p_1, p_2 and p_3 is an edge and ρ otherwise. As usual, here ρ denotes the edge density of G. This "strong" Laplacian seems to have intriguing potential but appears difficult to deal with.

The Laplacian and the eigenvalues of (2-) graphs have numerous applications ranging from extremal graph theory to randomized algorithms and approximation algorithms. Hypergraphs are general structures with rich properties. The Laplacian of a hypergraph is not only interesting on its own right but is also related to various applications such as amplifying random bits [35], communication complexity [13] and computational complexity. Basically, a boolean function can be viewed as an (ordered) hypergraph and a hypergraph can be viewed as a symmetric boolean function. Therefore, problems in various areas of computational complexity can perhaps be examined by using the Laplacian to capture the underlying structural properties.

9. Acknowledgement

The author is deeply indebted to Professor Charles Fefferman for his invaluable guidance. Thanks are also due to Peter Doyle, Phil Hanlon and Ron Graham for many helpful and illuminating discussions.

REFERENCES

1. N. Alon, Eigenvalues and expanders, *Combinatorica* 6 (1986) 83-86.
2. F. Bien, Constructions of telephone networks by group representations, *Notices Amer. Math. Soc.*, 36 (1989) 5-22.
3. Marianna Bolla, Spectra, Euclidean representations and vertex-colourings of hypergraphs, preprint.
4. Andrei Broder and Eli Shamir, On the second eigenvalue of random d-regular graphs, The 28th Annual Symposium on Foundations of Computer Science, (1987) 286-294.
5. F. R. K. Chung, Quasi-random classes of hypergraphs, *Random Structures and Algorithms* 1 (1990) 363-382.
6. F. R. K. Chung, Diameters and eigenvalues, *J. of Amer. Math. Soc.* 2 (1989) 187-196.
7. F. R. K. Chung, Constructing random-like graphs, AMS Short Course Lecture Notes 1991.
8. F. R. K. Chung and R. L. Graham, Quasi-random set systems, *J. of AMS*,4 (1991) 151-196.
9. F. R. K. Chung and R. L. Graham, Quasi-random subsets of Z_n, (to appear in *JCT(A)*).
10. F. R. K. Chung and R. L. Graham, Cohomological aspects of hypergraphs, (to appear in *TAMS*).
11. F. R. K. Chung, R. L. Graham and R. M. Wilson, Quasi-random graphs, *Combinatorica*, 9 (1989) 345-362.
12. F. R. K. Chung and R. L. Graham, Quasi-random hypergraphs,*Random Structures and Algorithms,* 1 (1990) 105-124.
13. F. R. K. Chung and P. Tetali, Communication complexity and quasi-randomness, (to appear in *SIAM J. Discrete Math.*)
14. D. Cvetković, M. Doob, I. Gutman, and A. Torgasev, Recent results in the Theory of Graph Spectra, North Holland (1988).
15. J. Dodzik and L. Karp, Spectral and function theory for combinatorial Laplacians, Geometry of Random Motion, *Contemp. Math* 73, AMS Publication (1988), 25-40.
16. Joel Friedman, On the second eigenvalue and random walks in random d-regular graphs, preprint.
17. G. Frobenius, Uber Matrizen aus nicht negative Elementen, Sitzber. Akad. Wiss. Berlin (1912) 456-477.
18. M. Fiedler, Algebraic connectivity of graphs, *Czech. Math. J.* 23 (1973), 298-305.

19. J. Friedman, J. Kahn and E. Szemeredi, On the second eigenvalues of random graphs, STOC (1989) 587-598.
20. J. Friedman and Avi Wigderson, On the second eigenvalue of hypergraphs, preprint.
21. F. R. Gantmacher, The Theory of Matrices, Vol. 1, Chelsea Pub. Co., New York (1977).
22. R. Grone, R. Merris, Coalescence majorization, edge valuations and the Laplacian spectra of graphs, *Linear and Multilinear Algebra* 27 (1990) 139-146.
23. K. Ireland and M. Rosen, A Classical Introduction to Modern Number Theory, Springer-Verlag, New York (1982).
24. F. Juhàsz, On the spectrum of a random graph, Colloq. Math. Soc. Janos Bolyai 25, Algebraic Methods in Graph Theory, Szeged (1978) 313-316.
25. N.M. Katz, An estimate for character sums, *J. Amer. Math. Soc.* 2 (1989) 197-200.
26. R. Lidl and H. Niederreiter, Finite Fields, Encyclopedia of Mathematics and its Applications, Cambridge Univ. Press, 1989.
27. A. Lubotsky, R. Phillips and P. Sarnak, Ramanujan graphs, *Combinatorica*, 8 (1988) 261-278.
28. G. A. Margulis, Explicit constructions of concentrators, *Problemy Peredaci Informacii*, 9 (1973) 71-80 (English transl. in *Problems Inform. Transmission*, 9 (1975) 325-332.
29. B. Mohar, Isoperimetric number of graphs, *J. Combin. Theory (B)* 47 (1989) 174-291.
30. J. R. Munkres, Elements of Algebraic Topology, Addison-Wesley, Reading Massachusetts, 1984.
31. O. Perron, Zur Theorie der Matrizen, *Math. Ann.*, 64 (1907) 248-263.
32. M. Pinsker, On the complexity of a concentrator, 7th Internat. Teletraffic Conf. Stockholm (1973) 318/1-4.
33. J. P. Serre, Linear Representations of Finite Groups, Springer-Verlag, New York (1977).
34. A. J. Sinclair and M.R. Jerrum, Approximate counting, uniform generation and rapidly mixing markov chain, to appear in Information and Computation.
35. M. Sipser, Expanders, randomness or time versus space, Structure and Complexity Theory (1986).
36. R. M. Tanner, Explicit construction of concentraters from generalized N-gons, *SIAM J. Algebraic and Discrete Methods* 5 (1984) 287-294.
37. R. M. Wilson, The necessary conditions for t-designs are sufficient, *Utilitas Math.* 4 (1973) 207-215.

BELLCORE, MORRISTOWN, NJ 07962
E-mail address: frkc@bellcore.com

Uniform sampling modulo a group of symmetries using Markov chain simulation

MARK JERRUM

ABSTRACT. Let Σ be a finite alphabet, and G a permutation group of degree n. The group G induces a natural action on Σ^n by permutation of the positions of symbols. This action partitions Σ^n into orbits, i.e., subsets of Σ^n which are equivalent modulo the symmetry group G. The problem addressed is that of uniform sampling from the set of all orbits. This problem is closely allied to that of estimating the cycle index polynomial of G, and can be regarded as part of the wider question "to what extent can Pólya theory be automated?". The approach taken here is to simulate a Markov chain on Σ^n that converges to a uniform distribution on the orbits. The Markov chain may be rapidly mixing for all G, but positive results to date are limited to some very simple special cases. The general question is likely to be difficult to resolve, because it includes the mixing rate of the Swendsen-Wang process as a special case.

1. Problem description

Let Σ be a finite alphabet of cardinality k, and G be a permutation group on $[n] = \{0, \ldots, n-1\}$. The group G has a natural action on the set Σ^n of all words of length n over the alphabet Σ, under which the word $x = x_0 x_1 \ldots x_{n-1}$ is mapped by the permutation $g \in G$ to the word $x^g = y_0 y_1 \ldots y_{n-1}$ defined by $y_j = x_i$ for all $i, j \in [n]$ satisfying $i^g = j$.[1] The action of G partitions Σ^n into a number of *orbits*, these being the equivalence classes of Σ^n under the equivalence relation that identifies x and y whenever there exists $g \in G$ mapping x to y.

1991 *Mathematics Subject Classification.* Primary 68Q25; Secondary 60J15, 60J20, 20B40.

The author was supported in part by grant GR/F 90363 of the UK Science and Engineering Research Council.

The final version of this paper will be submitted for publication elsewhere.

[1] Note that the permutation g is considered to act on *positions* rather than *indices*, since this is perhaps the easier to grasp of the two possible conventions.

By way of example, let $\Sigma = \{0, 1\}$, and $n = m^2$ for some integer m. Interpret each element of Σ^n as the adjacency matrix of an m-vertex directed graph Γ. Let G be the permutation group of of degree n and order $m!$ whose elements correspond to permutations of the vertex set of Γ, i.e., to simultaneous permutations of the rows and columns of the adjacency matrix. Then the orbits of Σ^n under the action of G correspond naturally to unlabelled directed graphs on m vertices. Many other sets of unlabelled combinatorial structures can be obtained using a like construction.

What has just been described is the setting for *Pólya theory*, the key result of which is an expression for the number of orbits in terms of the *cycle index polynomial* of G [2]. This polynomial, in the variables $z_1, z_2, \ldots, z_n$, is defined to be

$$P_G(z_1, \ldots, z_n) = |G|^{-1} \sum_{g \in G} z_1^{c_1(g)} \ldots z_n^{c_n(g)},$$

where c_i denotes the number of cycles in g of length i. The key result just referred to is that the number of orbits of Σ^n under the action of G is P_G evaluated at the point $z_1 = \cdots = z_n = k$. For many important choices for G, this computation is feasible and leads to results concerning the number of unlabelled combinatorial structures of various kinds [7].

For general groups G it seems that P_G is hard to compute exactly. Indeed, Goldberg has shown that evaluating $P_G(2, \ldots, 2)$ is #P-complete,[2] even when G is an abelian 2-group [6]. (See also a related result of Lubiw [17].) However, the question of whether there is an efficient *approximation* algorithm for P_G remains open. This is one of a group of three related questions which form the main subject of this article. They are as follows.

(a) Is there a fully polynomial randomised approximation scheme (fpras) [15] for estimating $P_G(k, \ldots, k)$? That is to say, is there a randomised algorithm that takes as input a group G and $\epsilon > 0$, and produces as output a number Y (a random variable) such that

$$\Pr\left((1 - \epsilon)P_G(k, \ldots, k) \leq Y \leq (1 + \epsilon)P_G(k, \ldots, k)\right) \geq \tfrac{3}{4},$$

and, moreover, does so within time $\text{poly}(n, \epsilon^{-1})$?

(b) Is there a polynomial-time almost uniform sampler[3] [14] for the orbits of Σ^n under the action of G? That is to say, is there a randomised algorithm that takes as input a group G and $\epsilon > 0$, and produces as output a word $Y \in \Sigma^n$ (a random variable) such that for each orbit Δ,

$$(1 - \epsilon)N^{-1} \leq \Pr(Y \in \Delta) \leq (1 + \epsilon)N^{-1},$$

---

[2]The class #P is the analogue for counting problems of the more familiar class NP which arises in the classification of decision problems. A #P-complete problem is computationally as difficult as counting the number of satisfying assignments to a Boolean formula, or the number of accepting computations of a non-deterministic Turing machine.

[3]When this concept was first introduced, "generator" was used in place of "sampler", but the latter word is more specific.

where $N = P_G(k, \ldots, k)$ is the total number of orbits? The execution time is required to be bounded by $\text{poly}(n, \epsilon^{-1})$.

(c) Is there a polynomial-time "almost-w" sampler for G, where the weight function $w : G \to \mathbb{Z}$ is defined by $w(g) = k^{c(g)}$, where $c(g)$ denotes the number of cycles in g. That is to say, is there a randomised algorithm that takes as input a group G and $\epsilon > 0$, and produces as output a permutation $Y \in G$ (a random variable) such that for each $g \in G$,

$$(1-\epsilon)w(g)Z^{-1} \leq \Pr(Y = g) \leq (1+\epsilon)w(g)Z^{-1},$$

where $Z = |G|\, P_G(k, \ldots, k)$? Again, the execution time is required to be bounded by $\text{poly}(n, \epsilon^{-1})$.

It may be observed that we have chosen to use the degree n as a simple measure of the "input size" of the permutation group G, and the reader may question whether this is reasonable. The convention may be justified by noting, as Sims did [20], that every permutation group can be specified by *strong generating set* of $O(n^2)$ permutations.[4] Thus, any permutation group may be described using a number of bits which is polynomial in n. Furthermore, Furst, Hopcroft, and Luks demonstrated that many permutation group operations — of which the most basic is membership testing — can be performed in time polynomial in n, given a strong generating set for the group in question [5]. These observations confirm that the degree is a reasonable measure of the "size" of a permutation group in the current context. While we are on the subject, it should be noted also that the alphabet Σ, though arbitrary, is regarded as fixed, and does not form part of the problem instance.

It is worth recording at this point the curious fact that although question (a) as stated is open, the answer is known to be "no" (assuming RP $\neq$ NP) if k is allowed to be fractional [6]. It remains an intriguing possibility that the particular combinatorial significance of the cycle index polynomial when k is integral may provide a path to resolving question (a) in the affirmative, while avoiding the problems that arise in the case of fractional k.

The complexity of approximate counting and of almost uniform sampling are known to be closely related,[5] which would lead one to suppose that questions (a), (b), and (c) ought to be equivalent. However, the situation here is atypical, and it is not clear, for example, whether resolving question (a) in the affirmative would immediately settle either of the others. The two known entailments are described in the following proposition, whose proof — being rather tangential to the main theme of the article — is relegated to an appendix.

PROPOSITION 1.1. *An affirmative answer to question* (c) *would entail affirmative answers to* (a) *and* (b).

---

[4]Indeed, $O(n)$ permutations suffice [10].

[5]A rather precise statement of this relationship has been formulated by Jerrum, Valiant, and Vazirani [14].

This article will address questions (b) and (c), and hence indirectly (a), using the by now familiar technique of Markov chain simulation.

2. A Markov chain with the appropriate distribution

Our approach to sampling orbits is to simulate an appropriately defined Markov chain. This technique has proved fruitful on a number of occasions in recent years; previous applications include an algorithm of Jerrum and Sinclair for estimating the permanent of a 0,1-matrix [1, 11, 19], and one of Dyer, Frieze, and Kannan for estimating the volume of a convex body in n-dimensional space [4]. In this instance we wish to construct a Markov chain $\mathfrak{M} = \mathfrak{M}(G, \Sigma)$ whose state space is Σ^n and whose stationary distribution assigns equal probability to each orbit. In fact, we shall aim at something stronger, namely, a stationary distribution that assigns to each word $x \in \Sigma^n$ a probability inversely proportional to the size of the orbit x^G containing x.

The transition probabilities from a state $x \in \Sigma^n$ are specified by the following conceptually simple two-step experiment.

(1) Choose g uniformly at random (u.a.r.) from the point stabiliser $G_x = \{g \in G : x^g = x\}$.
(2) Choose y u.a.r. from the set $\{y \in \Sigma^n : y^g = y\}$.

The new state is y. Before analysing the stationary distribution of $\mathfrak{M}$, we should pause to consider the computational complexity of implementing the above experiment. Step (2) is computationally undemanding, and amounts to assigning, u.a.r. and independently, a symbol from Σ to each cycle of g. Step (1) is more interesting, and is equivalent (under randomised polynomial-time reducibility) to computing a setwise stabiliser in a permutation group, a task that includes deciding isomorphism of two graphs as a special case. The computational complexity of the setwise stabiliser problem is open: it is one of the very rare natural candidates for a problem that is in the class NP, but is neither in P nor NP-complete.

Although no general polynomial-time algorithm for implementing step (1) is known, and it is perfectly possible that none exists, there are significant classes of groups G for which step (1) does have an efficient implementation. Luks has shown that p-groups — groups in which every element has order a power of p for some prime p — is an example of such a class [18]. We shall see in Section 4 that the Markov chain $\mathfrak{M}$ is already interesting in the context of abelian 2-groups.

Returning to the Markov chain itself, we note immediately that $\mathfrak{M}$ is ergodic, since every state can be reached from every other in a single transition, by selecting the identity permutation in step (1). The easiest way to get at the stationary distribution is perhaps by considering a random walk on the bipartite graph B that has vertex bipartition (Σ^n, G) and edge set $\{(x, g) : x^g = x\}$. It is clear that the Markov chain $\mathfrak{M}$ can be viewed as sampling the random walk on B after every even step. Let $\pi : \Sigma^n \to [0, 1]$ denote the stationary distribution of $\mathfrak{M}$. Then $\pi(x)$ is proportional to the degree of vertex x in the graph B, which is $|G_x|$. It is an elementary group-theoretic fact that $|G_x| \times |x^G| = |G|$, and hence $\pi(x)$ is inversely proportional to $|x^G|$. We have therefore established:

THEOREM 2.1. *Let π be the stationary distribution of Markov chain $\mathfrak{M}$. Then $\pi(x) = |x^G|^{-1} P_G(k, \ldots, k)^{-1}$ for all $x \in \Sigma^n$; in particular, π assigns equal probability to each orbit x^G.* □

Dually we might consider the Markov chain $\mathfrak{M}'$ with state space G and transition probabilities modelled by an experiment in which steps (1) and (2) appear transposed. By relating $\mathfrak{M}'$ to the random walk on B we easily obtain:

THEOREM 2.2. *Let π' be the stationary distribution of Markov chain $\mathfrak{M}'$. Then $\pi'(g) = k^{c(g)} |G|^{-1} P_G(k, \ldots, k)^{-1}$ for all $g \in G$, where $c(g)$ denotes the number of cycles in the permutation g.* □

Note that the stationary distributions of the Markov chains $\mathfrak{M}$ and $\mathfrak{M}'$ match the sampling distributions specified in questions (b) and (c) of the previous section.

3. The mixing rate

We have seen that the stationary distribution of the Markov chain $\mathfrak{M}$ is appropriate for sampling u.a.r. from the set of orbits of the G-action on Σ^n. For the Markov chain simulation approach to be computationally efficient, it is necessary for $\mathfrak{M}$ to be *rapidly mixing*. In the theoretical computer science tradition, we give a precise meaning to this informal requirement by insisting that $\mathfrak{M}$ should be "close" to stationarity after a number of steps that is bounded by a polynomial in n. Since the size of the state space is exponential in n, this is a non-trivial requirement.

There are a number of ways of quantifying "closeness" to stationarity, but they are all essentially equivalent in this application. Consider an ergodic Markov chain with state space Ω. Let $x \in \Omega$ be an arbitrary state, and denote by $P^t(x, \cdot)$ the distribution of the state at time t given that x is the initial state. Let π be the stationary distribution of $\mathfrak{M}$. Then the *variation distance* at time t with respect to the initial state x is defined to be

$$\Delta_x(t) = \max_{S \subseteq \Omega} |P^t(x, S) - \pi(S)| = \tfrac{1}{2} \sum_{y \in \Omega} |P^t(x, y) - \pi(y)|.$$

The rate of convergence may be measured by the function

$$\tau_x(\varepsilon) = \min\{t : \Delta_x(t') \leq \varepsilon \text{ for all } t' \geq t\}.$$

The Markov chain simulation technique will yield an efficient almost uniform sampler for orbits of the G-action on Σ^n provided the Markov chain $\mathfrak{M}(G, \Sigma)$ mixes rapidly. In recent years, techniques have been developed for bounding the mixing rate of combinatorially defined Markov chains, using, among other ideas, the relation between mixing rate and the expansion properties of the Markov chain viewed as a graph. Sinclair has provided a useful survey of these techniques, in addition to presenting some sharpened bounds [21]. Nevertheless, proofs of rapid mixing still tend to be technically involved.

Since no counterexamples have been identified, it remains a possibility that for any fixed alphabet Σ, the Markov chain $\mathfrak{M}(G, \Sigma)$ is rapidly mixing for all choices of G; specifically, that there is a *fixed* polynomial (in n and $\log \varepsilon^{-1}$) that uniformly bounds $\tau_x(\varepsilon)$ for all possible groups G. The author would not hazard a guess as to the likelihood of this state of affairs. The following section provides evidence that even if a general result of this form is true, it is likely to be difficult to prove. A more realistic programme for the short term is to produce a catalogue of families of groups G for which rapid mixing of $\mathfrak{M}(G, \Sigma)$ can be rigorously demonstrated. The full version of the paper will make a tentative start on this task.

4. Encoding a ferromagnetic Ising system

The *Ising model* has been the focus of much attention in the physics and mathematics communities since it was first introduced by Lenz [16] and Ising [8] in the early 1920s. A detailed account of the Ising model and its history will not be presented here; a very readable survey is given by Cipra [3], while Welsh [23] considers the Ising model from a complexity-theoretic perspective and sets it in the context of other combinatorial problems in statistical physics.

Our aim in this section to to exhibit a reduction from sampling configurations of a ferromagnetic Ising system to sampling binary words modulo an appropriate group of symmetries. The fact that the group in question is an abelian 2-group suggests that the mixing rate of $\mathfrak{M}(G, \Sigma)$ may be of substantial interest, even for restricted classes of groups. An unexpected feature of the reduction is that the dual Markov chain $\mathfrak{M}'$ which results can be directly identified with the Swendsen-Wang process, a recently developed and ingenious approach to Monte Carlo simulation of the Ising model. Although experiments seem to suggest that the mixing rate of the Swendsen-Wang process is rapid, no rigorous bounds have so far been obtained. This failure provides a salutary warning that the programme outlined in the previous section may be a difficult one. If nothing else, the reduction which follows offers an interesting (if quirky) alternative view of the Swendsen-Wang process.

Let $\Gamma = (V, E)$ be an undirected graph defining an Ising system with sites V and bonds $E \subseteq V^{(2)}$, and let J be the interaction energy, assumed positive (i.e., the system is ferromagnetic) and constant over bonds. We assume throughout that the applied field is zero. We construct a permutation group G over the set $X = \bigcup_{e \in E} X_e$, which is the disjoint union of two-element sets X_e. For $e \in E$ and $v \in V$, denote by h_e the permutation that transposes the elements of X_e and leaves everything else fixed, and denote by g_v the generator $\prod_{e \ni v} h_e$. Finally, define $G = G(\Gamma) = \langle g_v : v \in V \rangle$.

Observe that the generators of G commute and have order two, so each permutation $g \in G$ can be expressed as

(4.1) $$g = g(\sigma) = \prod_{v \in V} g_v^{[1-\sigma(v)]/2} = \prod_{e=\{u,v\} \in E} h_e^{[1-\sigma(u)\sigma(v)]/2},$$

where $\sigma : V \to \{-1,+1\}$. (Note that all the exponents are either 0 or 1.) Provided the graph G is connected, expression (4.1) is essentially canonical, in that σ is uniquely determined modulo uniform inversion of signs. We shall resolve this remaining ambiguity by imposing the condition $\sigma(v_0) = +1$ for some distinguished vertex v_0.

The function σ can be interpreted as a normalised spin configuration of the Ising system, where the convention for labelling spins by ± 1 is chosen so that $\sigma(v_0) = +1$. (Since the applied field is zero, there is nothing to choose between the two possible labellings.) Note that the function mapping g to σ is a bijection between G and the set of normalised spin configurations.

Observe that, from expression (4.1), the number of cycles in a permutation $g = g(\sigma)$ may be written

$$c(g) = \sum_{e=\{u,v\} \in E} [3 + \sigma(u)\sigma(v)]/2.$$

The cycle index evaluation at $k = 2$ is then

$$P_G(2,\ldots,2) = |G|^{-1} \sum_{\sigma:\sigma(v_0)=+1} 2^{c(g(\sigma))}$$
$$= \tfrac{1}{2}|G|^{-1} \sum_\sigma \prod_{e=\{u,v\}} \exp_2\left\{[3 + \sigma(u)\sigma(v)]/2\right\}$$
$$= \tfrac{1}{2}|G|^{-1} (2\sqrt{2})^{|E|} \sum_\sigma \prod_{e=\{u,v\}} (\sqrt{2})^{\sigma(u)\sigma(v)},$$

which, ignoring the easily computed scaling factor $2|G|(2\sqrt{2})^{-|E|}$, is exactly the partition function of the Ising system with underlying graph Γ, and $\beta J = \ln \sqrt{2}$, where β is "inverse temperature" and J the interaction energy.

Using more exotic constructions for the group G it would be possible to obtain the partition function of the Ising system at values of βJ different from $\ln \sqrt{2}$. However, is is easier to use a two-stage reduction, in which we first reduce to an Ising system with $\beta J = \ln \sqrt{2}$, and then apply the construction described above. The first stage of the reduction is straightforward, as a weak bond can be simulated by a chain of stronger bonds placed in series, and a strong bond by a bundle of weaker bonds placed in parallel.[6] By combining these two constructions it is possible to simulate general interactions economically and with high precision.

It should be observed that there is a direct correspondence between terms of the cycle index expansion, and terms of the Ising partition function. As a result,

---

[6]This is a special case of the *k-stretch* and *k-thickening* operations that Jaeger, Vertigan, and Welsh define and use in a much more general setting [9].

it is possible to sample an Ising configuration from the appropriate (i.e., Gibbs) distribution by sampling a permutation $g \in G$ from the stationary distribution of the dual Markov chain $\mathfrak{M}'(G(\Gamma), \{0,1\})$, and interpreting g as an spin configuration as in (4.1). If the Markov chain were rapidly mixing, this technique would provide a direct and efficient procedure for sampling spin configurations. At present, the only known approach to constructing a polynomial-time sampler for spin configurations is very indirect, and based on the algorithm of Jerrum and Sinclair for estimating the partition function [12, 13].

Looking more closely into the Markov chain $\mathfrak{M}'(G(\Gamma), \{0,1\})$, we discover that it does not represent an entirely new approach to Monte Carlo simulation of the Ising model, but merely an alternative development of an existing approach. The *Swendsen-Wang process* is a Markov chain on spin configurations in which the transition probabilities from a configuration $\sigma : V \to \{-1, +1\}$ are modelled by the following two step experiment.

(1) Form a set $A \subseteq E$ using a sequence of $|E|$ trials. Initially, set $A = \emptyset$; then, for each edge (bond) $e = \{u, v\} \in E$ with $\sigma(u) = \sigma(v)$, add e to A with probability $1 - e^{-2\beta J}$.

(2) Compute the connected components of the subgraph (V, A) of Γ. For each connected component, choose a spin independently and u.a.r. from $\{-1, +1\}$ and assign the chosen spin to all vertices in the component.

It is not too difficult to show that the transition probabilities defined by this experiment are identical to those of the dual Markov chain $\mathfrak{M}'(G(\Gamma), \{0,1\})$. In fact the correspondence is even closer, in that steps (1) and (2) can be identified with the two steps that make up a single transition of $\mathfrak{M}'$. (Thus step (1) above corresponds to step (2) of Markov chain $\mathfrak{M}$ defined in Section 2, and vice versa.) To bring out this correspondence it is only necessary to explain the connection between the subgraphs $A \subseteq E$ and words $y \in \Sigma^n = \{0,1\}^n$ which arise as "auxiliary states" in the Swendsen-Wang process and in $\mathfrak{M}'$, respectively. (Here, $n = |X| = 2|E|$.) With each auxiliary state $y : X \to \{0, 1\}$ of $\mathfrak{M}'$, associate an auxiliary state

$$A = A(y) = \{e \in E : y \text{ is not constant on } X_e\}.$$

of the Swendsen-Wang process. This rule naturally partitions Σ^n into equivalence classes and associates with each equivalence class an edge set $A \subseteq E$. Under this correspondence, the individual steps (two steps per transition) of the respective processes can be compared. It is then a matter of straightforward computation to show that the two processes are identical at the level of steps as well as transitions.

Appendix

PROOF OF PROPOSITION 1.1. We first show that (c) $\Rightarrow$ (b). Suppose there exists a polynomial-time almost-w sampler for G. The following two-step procedure constitutes an almost uniform sampler for the orbits of the G-action on Σ^n:

(1) using the almost-w sampler with input parameters (G,ε), sample $g \in G$ with probability approximately proportional to the weight $w(g) = k^{c(g)}$; (2) assign independently and u.a.r. an element from Σ to each cycle of g.

The outcome of this procedure is a r.v. $Y \in \Sigma^n$ satisfying

$$(1-\varepsilon)|G_x|Z^{-1} \leq \Pr(Y=x) \leq (1+\varepsilon)|G_x|Z^{-1},$$

for all $x \in \Sigma^n$, where $G_x = \{g \in G : x^g = x\}$ denotes the stabiliser of x, and $Z = |G|\,P_G(k,\ldots,k)$. To see this, observe that the event $Y = x$ can occur only if a permutation $g \in G_x$ is selected in step (1). Thus there are $|G_x|$ ways to obtain the outcome $Y = x$, and each occurs with probability Z^{-1}. Employing the elementary group-theoretic fact that $|G_x| \times |x^G| = |G|$ we obtain, letting $N = P_G(k,\ldots,k)$,

$$\frac{1-\varepsilon}{N\,|x^G|} \leq \Pr(Y=x) \leq \frac{1+\varepsilon}{N\,|x^G|},$$

which demonstrates that the two-step procedure given above would provide a satisfactory affirmative answer to question (b).

We now turn to the entailment (c) $\Rightarrow$ (a). For $i = 0,\ldots,n$, let $G_i \leq G$ denote the pointwise stabiliser $G_i = \{g \in G : j^g = j \text{ for all } j < i\}$. Observe that $G = G_0 \geq G_1 \geq \cdots \geq G_n = \{1\}$, and $|G_{i-1}| \leq n|G_i|$ for $i = 1,\ldots,n$. Generating sets for the pointwise stabilisers G_i may be read off from the strong generating set for G [5], so their use creates no computational difficulties. Our approach to estimating $P_G(k,\ldots,k)$ is to express the value in question as a product

$$P_G(k,\ldots,k) = k^n \prod_{i=1}^{n} \frac{P_{G_{i-1}}(k,\ldots,k)}{P_{G_i}(k,\ldots,k)}$$

(where we have used the elementary fact that $P_{\{1\}}(k,\ldots,k) = k^n$), and to estimate each of the factors using the Monte Carlo method. This approach is standard in the area [4, 11, 12, 13].

To meet the requirements of an fpras for $P_G(k,\ldots,k)$, it is enough to obtain estimates for each of the factors that are correct to within relative error $1 \pm \varepsilon/2n$ with probability at least $1-(4n)^{-1}$. To avoid unnecessary complications, assume $\varepsilon < 1$. We employ an estimator for the reciprocal

$$r_i = \frac{P_{G_i}(k,\ldots,k)}{P_{G_{i-1}}(k,\ldots,k)}$$

of the ith factor that is defined by the following two-step experiment: (1) using the postulated almost-w sampler with input parameters $(G_{i-1}, \varepsilon/7n)$, sample $g \in G_{i-1}$ with probability approximately proportional to the weight $w(g) = k^{c(g)}$; (2) if $g \in G_i$ then return the value $|G_{i-1}|/|G_i|$, else return 0. Letting the outcome of this experiment be the random variable Y_i, we have

$$(1-\varepsilon/7n)r_i \leq \mathrm{Exp}(Y_i) \leq (1+\varepsilon/7n)r_i.$$

Observe that $\max Y_i = |G_{i-1}|/|G_i|$ and $r_i \geq 1$, the latter inequality being a consequence of the combinatorial interpretation of $P_G(k,\ldots,k)$, and the fact that G_i is a subgroup of G_{i-1}. Thus the ratio $\text{Exp}(Y_i{}^2)(\text{Exp}\,Y_i)^{-2}$ — which is the critical parameter governing the efficiency of the Monte Carlo method — can be bounded above as follows:

$$\frac{\text{Exp}(Y_i{}^2)}{(\text{Exp}\,Y_i)^2} \leq \frac{\max Y_i}{\text{Exp}\,Y_i} \leq \frac{|G_{i-1}|}{|G_i|} \leq \frac{6n}{7}.$$

It follows, by a standard result in the area of Monte Carlo estimation, that $\text{O}(\varepsilon^{-2} n^3 \log n^{-1})$ trials suffice to obtain an estimate for $\text{Exp}\,Y_i$ that is correct to within relative error $1 \pm \varepsilon/7n$ with probability at least $1 - (4n)^{-1}$, and hence an estimate for r_i that is correct to within relative error $1 \pm \varepsilon/3n$ with similar probability. A relative error of $1 \pm \varepsilon/3n$ in the reciprocals of the factors translates to a relative error of at most $1 \pm \varepsilon/2n$ in the factors themselves. □

Acknowledgement

I thank Leslie Goldberg for comments on an early draft of this article.

References

1. A. Z. Broder, *How hard is it to marry at random? (On the approximation of the permanent)*, Proceedings of the 18th ACM Symposium on Theory of Computing, ACM Press, 1986, pp. 50–58, Erratum in Proceedings of the 20th ACM Symposium on Theory of Computing.
2. N. G. de Bruijn, *Pólya's theory of counting*, Applied Combinatorial Mathematics (E. F. Beckenbach, ed.), John Wiley and Sons, 1964, pp. 144–184.
3. B. Cipra, *An introduction to the Ising model*, American Mathematical Monthly **94** (1987), 937–959.
4. M. Dyer, A. Frieze, and R. Kannan, *A random polynomial time algorithm for approximating the volume of convex bodies*, Proceedings of the 21st ACM Symposium on Theory of Computing, ACM Press, 1989, pp. 375–381.
5. M. Furst, J. Hopcroft, and E. Luks, *Polynomial time algorithms for permutation groups*, Proceedings of the 21st IEEE Symposium on Foundations of Computer Science, Computer Society Press, 1980, pp. 36–41.
6. L. A. Goldberg, *Automating Pólya theory: the computational complexity of the cycle index polynomial*, Information and Computation (to appear).
7. F. Harary and E. M. Palmer, *Graphical enumeration*, Academic Press, 1973.
8. E. Ising, *Beitrag zur Theorie des Ferromagnetismus*, Zeitschrift für Physik **31** (1925), 253–258.
9. F. Jaeger, D. L. Vertigan, and D. J. A. Welsh, *On the computational complexity of the Jones and Tutte polynomials*, Mathematical Proceedings of the Cambridge Philosophical Society **108** (1990), 35–53.
10. M. Jerrum, *A compact representation for permutation groups*, Journal of Algorithms **7** (1986), 60–78.
11. M. R. Jerrum and A. J. Sinclair, *Approximating the permanent*, SIAM Journal on Computing **18** (1989), 1149–1178.
12. M. Jerrum and A. Sinclair, *Polynomial-time approximation algorithms for the Ising model (Extended Abstract)*, Proceedings of the 17th EATCS International Colloquium on Automata Languages and Programming, Springer-Verlag, 1990, pp. 462–475.
13. M. Jerrum and A. Sinclair, *Polynomial-time approximation algorithms for the Ising model*, SIAM Journal on Computing (to appear).
14. M. R. Jerrum, L. G. Valiant and V. V. Vazirani, *Random generation of combinatorial structures from a uniform distribution*, Theoretical Computer Science **43** (1986), 169–188.

15. R. M. Karp and M. Luby, *Monte-Carlo algorithms for enumeration and reliability problems*, Proceedings of the 24th IEEE Symposium on Foundations of Computer Science, Computer Society Press, 1983, pp. 56–64.
16. W. Lenz, *Beitrag zum Verständnis der magnetischen Erscheinungen in festen Körpern*, Zeitschrift für Physik **21** (1920), 613–615.
17. A. Lubiw, *Some NP-complete problems similar to graph isomorphism*, SIAM Journal on Computing **10** (1981), 11–21.
18. E. M. Luks, *Isomorphism of graphs of bounded valence can be tested in polynomial time*, Journal of Computer and System Sciences **25** (1982), 42–65.
19. M. Mihail, *On coupling and the approximation of the permanent*, Information Processing Letters **30** (1989), 91–95.
20. C. C. Sims, *Computational methods in the study of permutation groups*, Computational Problems in Abstract Algebra (J. Leech, ed.), Pergamon Press, New York, 1970, pp. 169–183.
21. A. Sinclair, *Improved bounds for mixing rates of Markov chains and multicommodity flow*, Combinatorics, Probability and Computing (to appear).
22. R. H. Swendsen and J-S. Wang, *Nonuniversal critical dynamics in Monte Carlo simulations*, Physical Review Letters **58** (1987), 86–88.
23. D. J. A. Welsh, *The computational complexity of some classical problems from statistical physics*, Disorder in Physical Systems, Oxford University Press, 1990, pp. 307–321.

DEPARTMENT OF COMPUTER SCIENCE, UNIVERSITY OF EDINBURGH, THE KING'S BUILDINGS, EDINBURGH EH9 3JZ, UK

E-mail address: mrj@dcs.ed.ac.uk

ON THE SECOND EIGENVALUE AND LINEAR EXPANSION OF REGULAR GRAPHS

NABIL KAHALE

ABSTRACT. The spectral method is the best currently known technique to prove lower bounds on expansion. We improve this technique by showing that the expansion coefficient of a linear-sized subset of a k-regular graph G is at least $\frac{k}{2}\left(1 - \sqrt{1 - (4k-4)/\tilde{\lambda}^2}\right)^-$, where $\lambda_1(G)$ is the second largest eigenvalue of the graph and $\tilde{\lambda} = \max(\lambda_1(G), 2\sqrt{k-1})$. In particular, the linear expansion of Ramanujan graphs is at least $(k/2)^-$. This improves upon the best previously known lower bound of $3(k-2)/8$. We also show that the average degree of the induced subgraphs on linear-sized subsets of Ramanujan graphs is at most $(1 + \sqrt{k-1})^+$, improving upon the best previously known bound of $(2\sqrt{k-1})^+$. For any integer k such that $k-1$ is prime, we explicitly construct an infinite family of k-regular graphs G_n on n vertices whose linear expansion is $k/2$ and such that $\lambda_1(G_n) \leq (2 + o(1))\sqrt{k-1}$. Since the graphs G_n have asymptotically optimal second eigenvalue, this essentially shows that $(k/2)$ is the best bound one can obtain using the second eigenvalue method.

1. INTRODUCTION

Given an undirected k-regular graph $G = (V, E)$ and a subset X of V, we define the expansion of X to be the ratio $|N_G(X)|/|X|$, where $N_G(X) = \{w \in V : \exists v \in X, (v,w) \in E\}$ is the set of neighbors of X. Graphs whose all subsets of size lying in a given range have large expansion are called expander graphs.

Expander graphs are widely used in Computer Science, in areas ranging from parallel computation [2, 5, 14, 18, 22] to complexity theory and cryptography [1, 6, 11, 23]. The range of the subsets whose expansion is relevant and the magnitude of the expansion needed depends on the nature of the application. For example, in the design of the AKS sorting circuit, we use expanders of constant degree such that subsets of size at most $\epsilon|V|$ have expansion at least $(1-\epsilon)/\epsilon$, where ϵ is a fixed positive constant. The depth of the resulting network is proportionnal to the

---

1991 *Mathematics Subject Classification.* 05C35, 68R10.

Supported by the Defense Advanced Research Projects Agency under Contracts N00014-92-J-1799 and N00014-91-J-1698, the Air Force under Contract F49620-92-J-0125, and the Army under Contract DAAL-03-86-K-0171. Part of this work was done while the author was visiting Dimacs.

The final version of this paper will be submitted for publication elsewhere

This paper was based on "On the Second Eigenvalue and Linear Expansion of Regular Graphs" by Nabil Kahale, which appeared in the *33rd Annual Symposium on Foundations of Computer Science*, Pittsburgh, Pennsylvania; October 24–27, 1992; pp. 296–303. ©IEEE.

degree of the expander. In other applications, like the construction of non-blocking networks in [5], we need a family of fixed degree uneven bipartite expanders where the expansion of linear-sized subsets is greater than $k/2$. Indeed, such an expansion guarantees that a constant fraction of any linear-sized subset have *unique neighbors*.

It is known that random regular graphs are good expanders. For example, for any $\beta < k-1$, there exists a constant α such that, with high probability, all the subsets of a random k-regular graph of size at most αn have expansion at least β. However, the explicit construction of expander graphs is much more difficult. The best currently known method to calculate lower bounds on the expansion in polynomial time relies on analysing the second eigenvalue of the graph. It is known that all the eigenvalues of the adjacency matrix $A(G)$ of G are real. Let $\lambda_i(G)$ denote the i-th largest eigenvalue of G. We have $\lambda_0(G) = k$ and $\lambda(G) = \max(\lambda_1(G), |\lambda_{n-1}(G)|) \leq k$, with equality iff G is not connected or bipartite. Tanner [21] showed that, for any subset X of a k-regular graph

$$(1) \qquad |N_G(X)| \geq \frac{k^2|X|}{\lambda^2 + (k^2 - \lambda^2)\frac{|X|}{n}}$$

The right-hand side of Eq. 1 increases as λ decreases. However, since $\liminf \lambda(G_n) \geq 2\sqrt{k-1}$ for any family [3] of k-regular graphs G_n, the best asymptotic expansion coefficient one can get by Tanner's result is $\approx k/4$. This bound is achieved by Ramanujan graphs. By definition, a Ramanujan graph is a connected k-regular graph whose eigenvalues $\neq \pm k$ are at most $2\sqrt{k-1}$ in absolute value. Infinite families of Ramanujan graphs have been explicitly constructed in [15, 16] when $k - 1$ is prime. The linear expansion of Ramanujan graphs was recently [12] improved to $3(k-2)/8$.

We define the linear expansion of a family of k-regular graphs G_n on n vertices to be the best lower bound on the expansion of subsets of size up to αn, where α is an arbitrary small positive constant. Our aim is to calculate the best linear expansion one can prove using the second eigenvalue technique. In this paper, we prove that the expansion of linear subsets of a k-regular graph G is at least $\frac{k}{2}\left(1 - \sqrt{1 - (4k-4)/\tilde{\lambda}^2}\right)^-$. In particular, linear-sized subsets of Ramanujan graphs have expansion at least $(k/2)^-$. On the other hand, for any integer k such that $k-1$ is a prime congruent to 1 modulo 4, and for any function m of n such that $m = o(n)$, we explicitly construct an infinite family of k-regular graphs G_n on n vertices such that $\lambda(G_n) \leq (2+o(1))\sqrt{k-1}$ and G_n contains a subset of size $2m$ with expansion $k/2$. Since such a family has asymptotically optimal second eigenvalue, this essentially shows that $k/2$ is the best bound lower bound on the linear expansion one can obtain by the second eigenvalue method. However, it is still an open question whether there exists a family of Ramanujan graphs with linear expansion $k/2$. We also show that the average degree of the induced subgraphs on linear-sized subsets of a k-regular graph G is at most $\left(1 + \tilde{\lambda}/2 + \sqrt{\tilde{\lambda}^2/4 - (k-1)}\right)^+$. This bound is equal to $(1 + \sqrt{k-1})^+$ in the case of Ramanujan graphs, improving upon the best previously known bound [4] of $(2\sqrt{k-1})^+$. Finally, for many integers k, we establish the existence of an infinite family of graphs on n nodes with diameter $(2 + o(1))\log_{k-1} n$ and $\lambda \leq (2 + o(1))\sqrt{k-1}$. Given the bound [8, 19]

$$D(G) \leq \left\lfloor \frac{\cosh^{-1}(n-1)}{\cosh^{-1}(k/\lambda)} \right\rfloor + 1 \tag{2}$$

on the diameter of any graph, this essentially shows that the bound $(2+o(1))\log_{k-1} n$ is the best one can obtain by the second eigenvalue method. As a biproduct of our techniques, we obtain known bounds on the number of edges in an induced subgraph of a regular graph, we give a simple proof of Tanner's inequality, and we establish a lower bound of $2\sqrt{k-1}(1+O(\log_k^{-2} n))$ on the second eigenvalue of any k-regular graph. A previous lower bound of $2\sqrt{k-1}(1+O(\log_k^{-1} n))$ was proven in [17] and improved to $2\sqrt{k-1}(1+O(\log_k^{-2} n))$ in [10]. Our results provide an efficient way to test that the expansion of linear sized subsets of random graphs is at least $k/2 + O(k^{3/4}\log^{1/2} k)$. As an application of the improved expansion of Ramanujan graphs, we can build explicit selection networks of asymptotic size $(3+\epsilon)n\log_2 n$, for any $\epsilon > 0$, improving on the bound $6n\log_2 n$ that was previously known. Our results also show that Ramanujan graphs of degree at least 7 are extrovert graphs. Such graphs have been used [7] to solve the token distribution problem. Classical results [4] require the degree to be at least 15.

2. Notation, definitions, and background

Throughout the paper, $G = (V, E)$ will denote an undirected connected graph on a set V of vertices. If G is regular, it is easy to see that $|N_G(X)| \geq |X|$ for any subset X. Let $L^2(V)$ denote the set of real valued functions on V and $L_0^2(V) = \{f \in L^2(V); \sum_{v \in V} f(v) = 0\}$. As usual, we define the scalar product of two vectors f and g of $L^2(V)$ by

$$f \cdot g = \sum_{v \in V} f(v) g(v),$$

and the euclidean norm of a vector f by $||f|| = \sqrt{f \cdot f}$. We denote the adjacency matrix of G by $A(G)$, or simply by A if there is no risk of confusion. $A(G)$ is the 0-1 $n \times n$ matrix whose (i,j) entry is equal to 1 iff $(i,j) \in E$. If we consider $f \in L^2(V)$ as a row vector, we have

$$(Af)(v) = \sum_{(v,w) \in E} f(w)$$

A defines a self-adjoint operator since $\forall f, g \in L^2(V)$ we have

$$(Af) \cdot g = f \cdot (Ag) = \sum_{(v,w) \in E} f(v)g(w) \tag{3}$$

The *girth* of G, denoted by $c(G)$, is the length of the shortest cycle in G. For any subset W of V, we denote by χ_W the characteristic vector of W:

$$\chi_W(v) = \begin{cases} 1 & \text{if } v \in W \\ 0 & \text{otherwise} \end{cases}$$

We define $B_l(W)$ to be the set of nodes at distance at most l from W. For any matrix M with real eigenvalues, we denote by $\lambda_i(M)$ the i-th largest eigenvalue of M. We also denote $\lambda_i(A(G))$ by $\lambda_i(G)$. If X and Y are two subsets of a graph $G = (V, E)$, then $e(X, Y) = |\{(u,v) \in X \times Y \cap E\}|$. If $Z_0, \ldots, Z_t$ is a sequence of non-empty subsets of V, we denote by $R^{\{Z_0, Z_1, \ldots, Z_t\}}$ the set of real valued functions on the set $\{Z_0, Z_1, \ldots, Z_t\}$. We define $\Phi(Z_0, Z_1, \ldots, Z_t)$ to be the linear mapping in $R^{\{Z_0, Z_1, \ldots, Z_t\}}$ whose matrix in the canonical basis $(\chi_{\{Z_0\}}, \ldots, \chi_{\{Z_t\}})$ is the $(t+1, t+1)$ matrix with entry (i,j) equal to $e(Z_i, Z_j)/|Z_i|$. The matrix $\Phi(Z_0, Z_1, \ldots, Z_t)$

can be viewed as the adjacency matrix of the weighted directed graph on the set $\{Z_0, Z_1, \ldots, Z_t\}$ where the weight of the edge (Z_i, Z_j) is equal to the average number of neighbors that a node in Z_i has in Z_j.

If M is a real matrix with nonnegative entries, then its largest eigenvalue $\lambda_0(M)$ is real. If s is a vector with positive entries and μ is a real such that $Ms \leq \mu s$, then $\lambda_0(M) \leq \mu$. This holds also if only the off-diagonal entries of M are assumed to be nonnegative, since we can reduce to the previous case by adding a multiple of the identity to M. Similarly, the inequalities $Ms \geq \mu s$ and $Ms \neq \mu s$ imply $\lambda_0(M) > \mu$.

3. Lower bound on the expansion

Lemma 1. *Let $G = (V, E)$ be a graph and let $Z_0, \ldots, Z_t$ be a sequence of nonempty disjoint subsets of V. For $0 \leq i \leq t$, we have $\lambda_i(G) \geq \lambda_i(\Phi)$, where $\Phi = \Phi(Z_0, Z_1, \ldots, Z_t)$.*

Proof. Define the scalar product $<,>$ on $R^{\{Z_0,\ldots,Z_t\}}$ by

$$<r, s> = \sum_{i=0}^{t} |Z_i| r(Z_i) s(Z_i).$$

Let ψ be the linear mapping from $R^{\{Z_0,\ldots,Z_t\}}$ to $L^2(V)$ that mapps $\chi_{\{Z_i\}}$ to χ_{Z_i}.

Claim 1. *For any $r, s \in R^{\{Z_0,\ldots,Z_t\}}$, we have $\psi(r) \cdot \psi(s) = <r, s>$.*

Proof. Since both sides of the above equality are bilinear in r and s, it suffices to show that the equality holds when r and s are elements of the canonical basis of $R^{\{Z_0,\ldots,Z_t\}}$. But $\psi(\chi_{\{Z_i\}}) \cdot \psi(\chi_{\{Z_j\}}) = \chi_{Z_i} \cdot \chi_{Z_j} = \delta_{i,j} |Z_i| = <\chi_{\{Z_i\}}, \chi_{\{Z_j\}}>$. □

Claim 2. *For any $r, s \in R^{\{Z_0,\ldots,Z_t\}}$, we have $\psi(r) \cdot A\psi(s) = <r, \Phi s>$.*

Proof. The claim follows from bilinearity and the equalities $\psi(\chi_{\{Z_i\}}) \cdot A\psi(\chi_{\{Z_j\}}) = \chi_{Z_i} \cdot A\chi_{Z_j} = e(Z_i, Z_j) = <\chi_{\{Z_i\}}, \Phi \chi_{\{Z_j\}}>$. □

This implies in particular that Φ defines a self-adjoint operator with respect to the product $<,>$ and so the eigenvalues of Φ are real. Using the elementary theory of quadratic forms and the injectivity of ψ, it follows that

$$\begin{aligned}
\lambda_i(\Phi) &= \max_{L} \min_{r \in L - \{0\}} \frac{<r, \Phi r>}{<r, r>} \\
&= \max_{L} \min_{f \in \psi(L) - \{0\}} \frac{f \cdot Af}{\|f\|^2} \\
&\leq \max_{L'} \min_{f \in L' - \{0\}} \frac{f \cdot Af}{\|f\|^2} \\
&= \lambda_i(G),
\end{aligned}$$

where L and L' range respectively over the subspaces of $R^{\{Z_0,\ldots,Z_t\}}$ and $L^2(V)$ of dimension $i + 1$. □

Similarly, one can prove that $\lambda_{n-1-i}(G) \leq \lambda_{t-i}(\Phi)$ for $0 \leq i \leq t$, but we will not need this inequality in the proof of Theorem 1.

Lemma 2. *Let $G = (V, E)$ be a graph and $X_{-1} = \emptyset, X_0, X_1, \ldots, X_t$ be a sequence of subsets of V such that, for $0 \leq i \leq t-1$, the degree of any element of X_i is equal to k, $N_G(X_i) \subseteq X_{i+1}$ and $|X_{i-1}| < |X_i|$. If the eigenvalues of the adjacency matrix of G are $\delta_0, \delta_1, \ldots, \delta_t$, with $|\delta_0| \geq |\delta_1| \geq \cdots \geq |\delta_t|$, then $|\delta_i|$ is greater than or equal to the i-th largest eigenvalue of the matrix $M_{t+1}(k; \rho_0, \rho_1, \ldots, \rho_{t-1})$ equal to*

$$\begin{pmatrix} 0 & k & 0 & 0 & \cdots & 0 \\ \rho_0 & 0 & k-\rho_0 & 0 & \cdots & 0 \\ 0 & \rho_1 & 0 & k-\rho_1 & \ddots & \vdots \\ \vdots & \ddots & \ddots & \ddots & \ddots & 0 \\ 0 & \cdots & 0 & \rho_{t-2} & 0 & k-\rho_{t-2} \\ 0 & 0 & \cdots & 0 & \rho_{t-1} & 0 \end{pmatrix},$$

where $\rho_i = k \frac{|X_i| - |X_{i-1}|}{|X_{i+1}| - |X_{i-1}|}$.

Proof. Note that $X_i \subseteq N^2(X_i) \subseteq X_{i+2}$. Consider the cover graph G_c of G defined on $V_c = V \times \{0,1\}$ and where $((u,l),(v,m)) \in V_c \times V_c$ is an edge iff $(u,v) \in E$ and $l \neq m$. The adjacency matrix of G_c is the tensor product of $A(G)$ and $A(K_2)$, and so the eigenvalues of G_c are the pairwise products of the eigenvalues of $A(G)$ and $A(K_2)$, namely $\lambda_0, \ldots, \lambda_{n-1}$ and $-\lambda_0, \ldots, -\lambda_{n-1}$, hence $\lambda_i(G_c) = |\delta_i|$ for $0 \leq i \leq n-1$. For $0 \leq i \leq t$, let Y_i be the subset of V_c defined by $Y_i = X_i \times \{(i \bmod 2)\}$. Then, for $0 \leq i \leq t-1$, the degree of any element of Y_i is equal to k, $N_{G_c}(Y_i) \subseteq Y_{i+1}$ and $Y_{i-1} \neq Y_{i+1}$. Now, we apply lemma 1 to G_c and the subsets $Z_i = Y_i - Y_{i-2}$, for $0 \leq i \leq t$. The Z_i's are non-empty and disjoint since Y_i is a strict subset of Y_{i+2} and $Y_i \cap Y_j = \emptyset$ if $i \not\equiv j \pmod{2}$. Note that $|Y_i| = |X_i|$ and $|Z_i| = |X_i| - |X_{i-2}|$. For $0 \leq i \leq t-1$, we have $e(Z_i, Z_{i+1}) = e(Y_i - Y_{i-2}, Y_{i+1} - Y_{i-1}) = e(Y_i, Y_{i+1} - Y_{i-1}) - e(Y_{i-2}, Y_{i+1} - Y_{i-1})$. But $e(Y_{i-2}, Y_{i+1} - Y_{i-1}) = 0$ since $N(Y_{i-2}) \subseteq Y_{i-1}$. So $e(Z_i, Z_{i+1}) = e(Y_i, Y_{i+1}) - e(Y_i, Y_{i-1}) = k|Y_i| - k|Y_{i-1}| = k(|X_i| - |X_{i-1}|)$. Finally, $N(Z_i) \subseteq N(Y_i) \subseteq Y_{i+1} = Z_{i+1} \cup Y_{i-1}$. On the other hand, $N(Y_{i-3}) \subseteq Y_{i-2}$ and so there are no edges between Z_i and Y_{i-3} (by convention $Y_{-2} = Y_{-3} = \emptyset$.) Hence $N(Z_i) \subseteq Z_{i+1} \cup Z_{i-1}$ and $e(Z_i, Z_j) = 0$ if $|i - j| \neq 1$. Therefore $\Phi_{G_c}(Z_0, Z_1, \ldots, Z_t) = M_{t+1}(k; \rho_0, \rho_1, \ldots, \rho_{t-1})$. This concludes the proof. $\square$

Lemma 3. [12] *Let W be a subset of a k-regular graph G and H the subgraph induced on W. Then $\lambda_0(H) \leq \lambda_1(G) + (k - \lambda_1(G))|W|/n$.*

Lemma 4. *Let $G = (V, E)$ be a k-regular connected graph and $X_0, X_1, \ldots, X_t$ a sequence of non-empty subsets of V such that, for $0 \leq i \leq t-1$, we have $N_G(X_i) \subseteq X_{i+1}$. For $l \geq 1$, $\lambda_1(G) + 4k^l|X_t|/n$ is greater than the largest eigenvalue of the matrix $M_{t+l}(k; \rho_0, \rho_1, \ldots, \rho_{t-1}, \underbrace{1, \ldots, 1}_{l-1 \text{ times}})$.*

Proof. Since the largest eigenvalue of the matrix M_{t+l} is at most k, we can assume without loss of generality that $|X_t|k^{l-1} \leq n$. We recursively construct subsets X_i of G, for $t < i \leq t+l-1$, such that $N_G(X_{i-1}) \subseteq X_i$ and $\rho_{i-1} = 1$. Let $i \in \{t, \ldots, t+l-2\}$ and assume that we constructed $X_{t+1}, \ldots, X_i$ such that $\rho_j = 1$ for $t \leq j < i$. The condition $\rho_j = 1$ implies that $|X_{j+1}| = k|X_j| - (k-1)|X_{j-1}| \leq k|X_j|$, and so $|X_i| \leq k^{i-t}|X_t| \leq \frac{n}{k}$. On the other hand, $e(X_i, V - X_{i-1}) = e(X_i, V) - e(X_i, X_{i-1}) = k|X_i| - k|X_{i-1}|$ since G is k-regular and $N_G(X_{i-1}) \subseteq X_i$. Therefore,

$|N_G(X_i) - X_{i-1}| \leq k|X_i| - k|X_{i-1}|$. But $|V - X_{i-1}| = n - |X_{i-1}| \geq k|X_i| - k|X_{i-1}|$. Therefore, there exists a subset X_{i+1} of V containing $N_G(X_i)$ and such that $|X_{i+1} - X_{i-1}| = k|X_i| - k|X_{i-1}|$. Since G is k-regular connected and since $N_G(X_{i-1}) \subseteq X_i$ and $|X_i| < n/2$, we have $|X_{i-1}| < |X_i|$ and so ρ_i is well defined and equal to 1.

Similarly, $|X_{j-1}| < |X_j|$ for $0 \leq j < t$ (we define X_{-1} to be the empty set). Now, we apply lemma 2 to the induced graph H on $W = X_{t+l-2} \cup X_{t+l-1}$ and the subsets $X_0, \ldots, X_{t+l-1}$. From the theory of matrices with non-negative entries [20], we know that $\lambda_0(H)$ is the largest eigenvalue of H in absolute value. Hence, $\lambda_0(H) \geq \lambda_0(M_{t+l}(k; \rho_0, \rho_1, \ldots, \rho_{t-1}, 1, \ldots, 1))$. By lemma 3, however,

$$\lambda_0(H) \leq \lambda_1(G) + (k - \lambda_1(G))\frac{|W|}{n} < \lambda_1(G) + 4\frac{|X_t|k^l}{n},$$

since $|\lambda_1| < k$ and $|W| \leq 2k^{l-1}|X_t|$. $\square$

Lemma 5. *The eigenvalues of the matrix $M_{l+1}(k; \rho_0, 1, \ldots, 1)$ are $2\sqrt{k-1}\cosh\theta$, where θ ranges over the solutions (in the complex domain) to the equation*

$$((4k-4)\cosh^2\theta - k\rho_0)s_l(\theta) = 2(k - \rho_0)\cosh(\theta)s_{l-1}(\theta),$$

and $s_i(\theta)$ is the analytical extension of the function $\frac{\sinh(i\theta)}{\sinh(\theta)}$ over the domain of complex numbers. If the largest eigenvalue of the matrix $M_{l+1}(k; \rho_0, 1, \ldots, 1)$ is at most $2\sqrt{k-1}\cosh(\theta')$, with $\theta' \geq 0$, then $\rho_0 \leq (1 + e^{2\theta'})(1 + O(1/l))$.

Proof. For ease of notations, we assume that $\sinh(\theta) \neq 0$. The case $\sinh(\theta) = 0$ can be treated by replacing $\sinh(i\theta)$ by $s_i(\theta)$ in the proof. First, we note that any real number λ can be written as $2\sqrt{k-1}\cosh(\theta)$, where θ is a complex number. Such λ is an eigenvalue of $M_{l+1}(k; \rho_0, 1, \ldots, 1)$ with eigenvector $(r_0, \ldots, r_l)$ iff

(4) $\quad \lambda r_0 = k r_1$
(5) $\quad \lambda r_1 = \rho_0 r_0 + (k - \rho_0) r_2$
(6) $\quad \lambda r_i = r_{i-1} + (k-1)r_{i+1}$ for $2 \leq i \leq l - 1$
(7) $\quad \lambda r_l = r_{l-1}$

From Eqs. 6 and 7, we see that, up to a constant factor, $r_i = (k-1)^{-i/2}\sinh((l+1-i)\theta)$ for $i \geq 1$. Eqs. 4 and 5 imply that $\lambda^2 r_1 = \rho_0 k r_1 + \lambda(k - \rho_0)r_2$, which reduces to the equation in the lemma.

Claim 3. *If θ is a nonnegative real and $l \geq 1$, then*

(8) $$\frac{l-1}{l}e^{-\theta} \leq \frac{\sinh((l-1)\theta)}{\sinh(l\theta)} \leq e^{-\theta}.$$

Proof. To prove inequality 8, we observe that

$$e^{-\theta} - \frac{\sinh((l-1)\theta)}{\sinh(l\theta)} = \frac{e^{-(l-1)\theta} - e^{-(l+1)\theta}}{e^{l\theta} - e^{-l\theta}} = e^{-l\theta}\frac{\sinh(\theta)}{\sinh(l\theta)} \leq \frac{e^{-\theta}}{l},$$

since $\sinh(l\theta) \geq l\sinh(\theta)$. $\square$

By setting $h(\theta') = (\lambda'^2 - k\rho_0)\sinh(l\theta') - 2(k - \rho_0)\cosh\theta'\sinh((l-1)\theta')$, where $\lambda' = 2\sqrt{k-1}\cosh\theta'$, we see that $h(\theta')$ goes to $+\infty$ as θ' goes to $+\infty$, and so the condition in the lemma implies that $h(\theta') \geq 0$. Using Eq. 8, we get

$$(\lambda'^2 - k\rho_0) \geq 2(k - \rho_0)\cosh(\theta')\frac{\sinh((l-1)\theta')}{\sinh(l\theta')}$$
$$= 2(k - \rho_0)\cosh(\theta')e^{-\theta'}(1 + O(\frac{1}{l}))$$

Therefore,

(9) $$\rho_0(k - 2e^{-\theta'}\cosh\theta') \leq (\lambda'^2 - 2ke^{-\theta'}\cosh\theta')(1 + O(\frac{1}{l}))$$

since $e^{-\theta'}\cosh\theta' \leq 1$, $k \geq 3e^{-\theta'}\cosh\theta'$ and $\lambda'^2 \geq 8k/3$. By factoring $2\cosh\theta'$ in the righthand side of Eq. 9 and noting that

$$(2k-2)\cosh\theta' - ke^{-\theta'} = ke^{\theta'} - 2\cosh\theta',$$

we see that $\rho_0 \leq 2\cosh(\theta')e^{\theta'}(1 + O(1/l))$. □

Theorem 1. *If $G = (V, E)$ is k-regular and $\tilde{\lambda} = \max(\lambda_1(G), 2\sqrt{k-1})$, then for any non-empty subset X of size at most $k^{-1/\epsilon}|V|$,*

$$\frac{|N_G(X)|}{|X|} \geq \frac{k}{2}\left(1 - \sqrt{1 - \frac{4k-4}{\tilde{\lambda}^2}}\right)(1 + O(\epsilon)),$$

where the constant behind the O is a small absolute constant.

Proof. Let $\tilde{\lambda} = 2\sqrt{k-1}\cosh(\theta)$, where $\theta \geq 0$. We apply lemma 4 with $t = 1$, $X_0 = X$ and $X_1 = N_G(X)$ and $l = \lfloor 1/2\epsilon \rfloor$. Let $\lambda' = \tilde{\lambda} + 4k^l|X_t|/n = 2\sqrt{k-1}\cosh(\theta')$, where $\theta' \geq \theta \geq 0$. Since $|X_1| \leq k|X| \leq k^{1-1/\epsilon}n$, we have $|X_1|k^l \leq k^{1-1/2\epsilon}n$ and $\lambda' - \tilde{\lambda} \leq 4k^{1-1/2\epsilon} = O(\epsilon^2)$. This implies that $\lambda' = \tilde{\lambda}(1 + O(\epsilon))$ and $\cosh\theta' - \cosh\theta = O(\epsilon^2)$. Using the inequality $(x-y)^2 \leq 2(\cosh x - \cosh y)$ valid for $x \geq y \geq 0$, we see that $\theta' - \theta = O(\epsilon)$. Lemma 5 then implies

$$\rho_0 \leq (1 + e^{2\theta'})(1 + O(\epsilon)) = (1 + e^{2\theta})(1 + O(\epsilon)).$$

Hence
$$\frac{|N_G(X)|}{|X|} = \frac{k}{\rho_0} \geq \frac{k}{2e^\theta\cosh\theta}(1 + O(\epsilon)).$$

Using the equality $e^{-\theta} = \cosh\theta - \sqrt{\cosh^2\theta - 1}$, we get

$$\frac{1}{e^\theta\cosh\theta} = 1 - \sqrt{1 - \frac{1}{\cosh^2\theta}} = 1 - \sqrt{1 - \frac{4k-4}{\tilde{\lambda}^2}}$$

This concludes the proof. □

Theorem 2. *If $G = (V, E)$ is k-regular and $\tilde{\lambda} = \max(\lambda_1(G), 2\sqrt{k-1})$, then for any non-empty subset X of size at most $k^{-1/\epsilon}|V|$,*

$$\frac{|e(X, X)|}{|X|} \leq \left(1 + \frac{\tilde{\lambda}}{2} + \sqrt{\frac{\tilde{\lambda}^2}{4} - (k-1)}\right)(1 + O(\epsilon)),$$

where the constant behind the O is a small absolute constant.

Proof. Let $\tilde{\lambda} = 2\sqrt{k-1}\cosh(\theta)$, where $\theta \geq 0$. Let $l = \lfloor \frac{1}{2\epsilon} \rfloor$ and let X_i be the set of nodes at distance i from X. By applying lemma 1 to the graph H induced on $\bigcup_{i=0}^{l} X_i$, we see that $\lambda_0(H) \geq \lambda_0(\Phi(X_0, \ldots, X_l))$. But $\lambda_0(H) \leq \lambda'$, where $\lambda' = \tilde{\lambda} + 4k^{l+1}|X_t|/n = 2\sqrt{k-1}\cosh(\theta')$ and $\theta' \geq \theta \geq 0$. As in the proof of theorem 1, we have $\theta' - \theta = O(\epsilon)$. Now, let $r_i = \sinh((l+1-i)\theta')(k-1)^{-i/2}$, for $0 \leq i \leq l$. The sequence r_i is strictly positive and decreasing, and we have

(10) $\qquad \lambda' r_i = r_{i-1} + (k-1)r_{i+1}$ for $1 \leq i \leq l-1$

(11) $\qquad \lambda' r_l = r_{l-1}$

For $1 \leq i \leq l-1$, we have

$$(\Phi r)_i = \Phi_{i,i-1} r_{i-1} + \Phi_{i,i} r_i + \Phi_{i,i+1} r_{i+1} \geq r_{i-1} + (k-1) r_{i+1} = \lambda' r_i,$$

since $\Phi_{i,i-1} \geq 1$, $\Phi_{i,i-1} + \Phi_{i,i} + \Phi_{i,i+1} = k$ and the sequence r_i is decreasing. Similarly, $(\Phi r)_l \geq \lambda' r_l$. Since $\lambda_0(\Phi) \leq \lambda'$ and the sequence r_i is strictly positive, this implies that $(\Phi r)_0 \leq \lambda' r_0$. Since $(\Phi r)_0 = \Phi_{0,0} r_0 + (k - \Phi_{0,0}) r_1$, we successively obtain

$$\begin{aligned}
\Phi_{0,0} &\leq \frac{\lambda' r_0 - k r_1}{r_0 - r_1} \\
&= \frac{2(k-1)\cosh(\theta')e^{\theta'} - k}{\sqrt{k-1}e^{\theta'} - 1}(1 + O(\epsilon)) \\
&= (\sqrt{k-1}e^{\theta} + 1)(1 + O(\epsilon))
\end{aligned}$$

The last equation follows from Eq. 8 since

$$\frac{r_0}{r_1} = \sqrt{k-1}\frac{\sinh((l+1)\theta')}{\sinh(l\theta')} = \sqrt{k-1}e^{\theta'}(1 + O(\frac{1}{l}))$$

We conclude by replacing $\Phi_{0,0}$ by its value and noting that

$$e^\theta = \cosh\theta + \sqrt{\cosh^2\theta - 1}. \quad \square$$

Corollary 1. *If G is k-regular on n vertices, then $\lambda_1(G) \geq 2\sqrt{k-1}(1 + O(\log_k^{-2} n))$.*

Proof. We apply lemma 4 with X_0 consisting of a single vertex, $t = 0$ and $l = \lfloor (\log_k n)/2 \rfloor$. From the theory of non-negative matrices [20], we know that the largest eigenvalue of the matrix $M_l(k; 1, \ldots, 1)$ is no smaller than the largest eigenvalue of the same matrix where the $(1,2)$ entry is replaced by $k-1$. But a calculation similar to the one in lemma 5 shows that the largest eigenvalue of this matrix is $\cos(\frac{\pi}{l+1})$. Hence $\lambda_1(G) + 4k^l/n \geq 2\sqrt{k-1}\cos(\frac{\pi}{l+1})$. We conclude the proof by noting that $k^l/n = O(\log_k^{-2} n))$ and $\cos(\frac{\pi}{l+1}) = 1 + O(\log_k^{-2} n)$. $\quad \square$

As we mentionned in the introduction, the lower bound in Corollary 1 was independently obtained in [10].

4. A FAMILY OF "ALMOST" RAMANUJAN GRAPHS WITH EXPANSION $k/2$

Lemma 6. *Let $G = (V, E)$ be a graph, X a subset of V and X_i the set of nodes at distance i from X. Assume that for $0 \leq i, j \leq l$, all nodes in X_i have the same number of neighbors in X_j. If $\lambda > 0$ and $s \in L^2(V)$ has positive components, is constant on each X_i and such that $As \leq \lambda s$ on $B_{l-1}(X)$, then for any $g \in L^2(V)$ such that $|Ag(u)| = \lambda|g(u)|$ for $u \in B_{l-1}(X)$, the sequence*

$$\frac{\sum_{v \in X_i} g(v)^2}{\sum_{v \in X_i} s(v)^2}$$

is an increasing function of i, for $0 \leq i \leq l$.

Proof. Let P, P_{l-1} and P_l be the projections on the sets $B_{l-1}(X), X_{l-1}$ and X_l respectively. By induction, it suffices to show that $\|P_l g\|/\|P_l s\| \geq \|P_{l-1} g\|/\|P_{l-1} s\|$. Let A' be the adjacency matrix of the subgraph induced on $B_l(X)$. By the regularity conditions in the lemma, there exist positive coefficients α, β and γ be such that $P_l A' s = \gamma P_l s$ and $A' P_l s = \alpha P_l s + \beta P_{l-1} s$. By hypothesis, we have $A' s \leq \lambda P s + \gamma P_l s$. Premultiplying both sides of this equation by P and A' successively yields

$$\begin{aligned} PA's &\leq \lambda s - \lambda P_l s \\ A'PA's &\leq \lambda A's - \lambda(\alpha P_l s + \beta P_{l-1} s) \\ &\leq \lambda^2 Ps + \lambda(\gamma - \alpha)P_l s - \lambda \beta P_{l-1} s \end{aligned}$$

But the matrix $A'PA' - \lambda^2 P - \lambda(\gamma - \alpha)P_l + \lambda \beta P_{l-1}$ has only nonnegative entries off its diagonal, and so its largest eigenvalue is 0 since s is positive. The quadratic form associated to this matrix is therefore negative, and so

$$A'PA'g \cdot g \leq \lambda^2 Pg \cdot g + \lambda(\gamma - \alpha)P_l g \cdot g - \lambda \beta P_{l-1} g \cdot g$$

In other words, $\|PA'g\|^2 \leq \lambda^2 \|Pg\|^2 + \lambda(\gamma - \alpha)\|P_l g\|^2 - \lambda\beta\|P_{l-1}g\|^2$. But $\|PA'g\| = \lambda\|Pg\|$ by hypothesis, and so

$$(12) \qquad (\gamma - \alpha)\|P_l g\|^2 \geq \beta \|P_{l-1} g\|^2$$

On the other hand, since A' and P_l are symmetric, we have $A'P_l s \cdot s = P_l A' s \cdot s$, and so $\beta\|P_{l-1}s\|^2 + \alpha\|P_l s\|^2 = \gamma\|P_l s\|^2$. Comparing this with Eq. 12 concludes the proof. □

Lemma 7. *If $G = (V, E)$ is k-regular on n vertices, for any $f \in L^2(V)$, we have*

$$f \cdot Af \leq \lambda_1(G)\|f\|^2 + \frac{k}{n}\Big(\sum_{v \in V} f(v)\Big)^2$$

Proof. Let $\overline{f} = \frac{f \cdot \chi_V}{n}\chi_V$ be the orthogonal projection of f on the space spanned by the constant vector χ_V. Then $f_0 = f - \overline{f}$ is the orthogonal projection of f on $L_0^2(V)$. We have $Af = A\overline{f} + Af_0 = k\overline{f} + Af_0$, and so $f \cdot Af = k\|\overline{f}\|^2 + f_0 \cdot Af_0$ since $\overline{f} \cdot Af_0 = A\overline{f} \cdot f_0 = k\overline{f} \cdot f_0 = 0$. But $\|\overline{f}\|^2 = \frac{(f \cdot \chi_V)^2}{n^2}\|\chi_V\|^2 = \big(\sum_{v \in V} f(v)\big)^2/n$, and $f_0 \cdot Af_0 \leq \lambda_1(G)\|f_0\|^2 \leq \lambda_1(G)\|f\|^2$ since $\|f\|^2 = \|f_0\|^2 + \|\overline{f}\|^2$. □

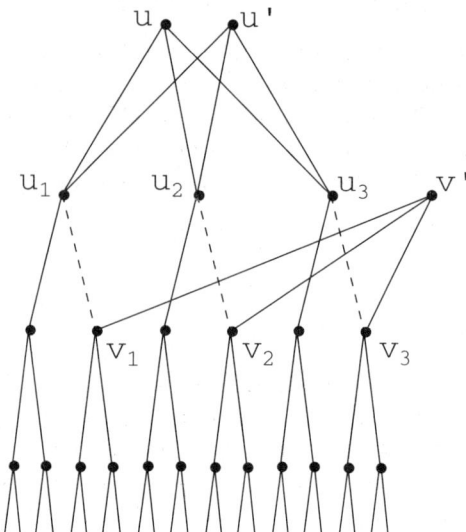

FIGURE 1. The graph G_{n+2} in the neighborhood of u in the case $k = 3$. The dotted edges are those belonging to $E - E'$.

Theorem 3. *For any integer k such that $k-1$ is prime, we can explicitly construct an infinite family of k-regular graphs G_n on n vertices whose linear expansion is $k/2$ and such that $\lambda_1(G_n) \leq 2\sqrt{k-1}(1 + 2\log^2 \log n / \log_k^2 n)$.*

Proof. From [15] and [16], we know that we can explicitly construct an infinite family of bipartite Ramanujan graphs H_n on n vertices whose girth is $(4/3 + o(1))\log_{k-1} n$. Let $H_n = (V, E)$ be an element of the family, $u \in V$ a vertex of H_n and $l = \lfloor c(H_n)/2 \rfloor - 2$. Let $u_1, \ldots, u_k$ be the neighbors of u and let $v_1, \ldots, v_k$ be k vertices at distance two from u such that $(u_i, v_i) \in E$. The subgraph of H_n induced on $B_{l+1}(\{u\})$ is a tree rooted at u since it is connected and contains no cycle. Let u' and v' be two elements not belonging to V. Consider the k-regular graph $G_{n+2} = (V', E')$, where $V' = V \cup \{u', v'\}$ and $E' = E \cup \bigcup_{i=1}^{k}\{(u', u_i), (u_i, u'), (v', v_i), (v_i, v')\} - \bigcup_{i=1}^{k}\{(u_i, v_i), (v_i, u_i)\}$. Figure 1 shows the graph G_{n+2} in the neighborhood of u in the case $k = 3$. For shorthand, we denote $A(G_{n+2})$ by A' and $\lambda_1(A')$ by λ_1. Assume that $\lambda_1 > 2\sqrt{k-1}$ (otherwise we are done), and let $\lambda_1 = 2\sqrt{k-1}\cosh\theta$, with $\theta > 0$. Let $g \in L_0^2(V')$ be an eigenvector corresponding to λ_1. Since u and u' have the same neighbors in G_{n+2} and $\lambda_1 \neq 0$, we have $g(u) = g(u')$. Let f be the element of $L^2(V)$ that coincides with g on V. By Eq. 3, we have

$$\lambda_1\|g\|^2 = g \cdot A'g$$
$$= f \cdot A(H_n)f - 2\sum_{i=1}^{k} g(u_i)g(v_i) + 2\sum_{i=1}^{k} g(u')g(u_i) + 2\sum_{i=1}^{k} g(v')g(v_i)$$
$$(13) \qquad \leq f \cdot A(H_n)f + \sum_{i=1}^{k}(g(u_i)^2 + g(v_i)^2) + 2\lambda_1 g(u')^2 + 2\lambda_1 g(v')^2$$

In the third inequality, we used the equation $(A'g)(u') = \lambda_1 g(u')$ and $(A'g)(v') = \lambda_1 g(v')$. Note that $\sum_{w \in V} f(w) = -g(u') - g(v')$ since $g \in L_0^2(V')$. Using lemma 7, we get

$$\begin{aligned} f \cdot A(H_n) f &\leq \lambda_1(H_n) \|f\|^2 + \frac{k}{n}(g(u') + g(v'))^2 \\ &\leq 2\sqrt{k-1}(\|g\|^2 - g(u')^2 - g(v')^2) + \frac{2k}{n}(g(u')^2 + g(v')^2) \\ &\leq 2\sqrt{k-1}\|g\|^2, \end{aligned}$$

for sufficiently large n. Combining this with Eq. 13, we obtain

$$\lambda_1 \|g\|^2 \leq 2\sqrt{k-1}\|g\|^2 + \sum_{i=1}^{k}(g(u_i)^2 + g(v_i)^2) + 2\lambda_1 g(u')^2 + 2\lambda_1 g(v')^2$$

(14) $$\leq 2\sqrt{k-1}\Big(\|g\|^2 + 2\sum_{i=1}^{k}(g(u_i)^2 + g(v_i)^2)\Big)$$

The second inequality follows from

$$g(u')^2 = \frac{1}{\lambda_1^2}\Big(\sum_{i=1}^{k} g(u_i)\Big)^2 \leq \frac{k}{\lambda_1^2}\sum_{i=1}^{k} g(u_i)^2$$

and a similar relation corresponding to v'. Let s be the function on V' defined by $s(u) = s(u') = k$ and, for $v \in V' - \{u, u'\}$ at distance i form u,

$$s(v) = 2(k-1)^{1-i/2} \cosh i\theta$$

An easy calculation shows that the function s verifies the conditions of lemma 6 with $\lambda = \lambda_1$, and so

$$\begin{aligned} \|g\|^2 &\geq \sum_{v \in X_{l+1}} g(v)^2 \\ &\geq \frac{\sum_{v \in X_{l+1}} s(v)^2}{\sum_{v \in X_1} s(v)^2} \sum_{v \in X_1} g(v)^2 \\ &\geq \frac{1}{2}\cosh^2(l\theta) \sum_{i=1}^{k} g(u_i)^2, \end{aligned}$$

since $\cosh(l+1)\theta \geq \cosh\theta \cosh l\theta$. Similarly, we can prove the inequality $\|g\|^2 \geq \frac{1}{2}\cosh^2(l\theta) \sum_{i=1}^{k} g(v_i)^2$. Combining this with Eq. 14 yields

(15) $$\cosh\theta \leq 1 + \frac{8}{\cosh^2 l\theta}$$

Using the inequalities $\cosh\theta \geq 1 + \theta^2/2$ and $\cosh l\theta \geq e^{l\theta}/2$, we see from Eq. 15 that $\theta = O(e^{-l\theta})$ and $\log\theta \leq -l\theta + O(1)$. Hence $\theta \leq (\log l)/l$ for sufficiently large n. Finally,

$$\lambda_1 = 2\sqrt{k-1}\cosh\theta = 2\sqrt{k-1}(1 + \frac{\theta^2}{2}(1+o(1))) \leq 2\sqrt{k-1}(1 + 2\frac{\log^2\log n}{\log_k^2 n}),$$

since $l \geq (2/3 + o(1))\log_k n$. Theorem 1 implies that the linear expansion of the family G_n is at least $k/2$. Since the subset $\{u, u'\}$ has k neighbors, this bound is tight. □

If $k - 1$ is a prime congruent to 1 modulo 4, we know from [15] that there exists an infinite family of non-bipartite k-regular Ramanujan graphs with girth at least $(2/3 + o(1))\log_{k-1} n$. By doing the same construction as in Theorem 3, we obtain k-regular graphs whose second largest eigenvalue *in absolute value* is $(2 + o(1))\sqrt{k-1}$ and linear expansion $k/2$. Moreover, by adding nodes at regions of the graph at sufficiently large distance from each other, we can construct for any $m = m(n) = o(n)$ a family of k-regular graphs whose second largest eigenvalue in absolute value is $(2+o(1))\sqrt{k-1}$ and containing a subset of size $2m$ with expansion $k/2$. The proof of these two statements is very similar to the proof of Theorem 3. We also have the following theorem.

Theorem 4. *For any integer k such that $k - 1$ is prime congruent to 1 modulo 4, there exists an infinite family of k-regular graphs on n vertices with $\lambda(G_n) = (2 + o(1))\sqrt{k-1}$ and diameter $(2 + o(1))\log_{k-1} n$.*

Proof. Sketch. We start with a k-regular Ramanujan graph H on n' vertices, and we form two identical trees of depth $\lfloor \log_{k-1} m - 2 \rfloor$, where $m = \lfloor n'/\log n' \rfloor$, and whose internal nodes have degree k. We identify the leaves of the two trees with nodes of H at distance $\Omega(\log_{k-1}(n'/m))$ from each other in H. As in Theorem 3, we delete $O(m)$ edges and add $O(m)$ edges and nodes to obtain a k-regular graph G_n. Using similar techniques to the proof of Theorem 3, we can show that $\lambda(G_n) = 2\sqrt{k-1}(1 + o(1))$. Eq. 2 implies that the diameter of G_n is at most $(2 + o(1))\log_{k-1} n$. On the other hand, it is at least twice the depth of the trees, which is $(1 + o(1))\log_{k-1} n$. □

5. EXAMPLES AND APPLICATIONS

(1) **Number of edges in an induced subgraph.** Let $G = (V, E)$ be a k-regular graph on n vertices and Z_0 a proper subset of V. We apply lemma 1 and the remark following it with $t = 1$ and $Z_1 = V - Z_0$. An easy calculation shows that $\lambda_0(G) = \lambda_0(\Phi) = k$ and $|Z_0||Z_1|\lambda_1(\Phi) = ne(Z_0, Z_0) - k|Z_0|^2$. The inequality $|\lambda_1(\Phi)| \leq \lambda$ then becomes $|e(Z_0, Z_0) - k|Z_0|^2/n| \leq \lambda|Z_0|(1 - |Z_0|/n)$. Note that this inequality is also true in the case $Z_0 = \emptyset$ or $Z_0 = V$. This result has already been established in [4].

(2) **Tanner's inequality.** Again, we assume that $G = (V, E)$ is a k-regular graph on n vertices. Let X be a proper subset of V. We apply lemma 2 with $t = 3$, $X_0 = X$, $X_1 = N_G(X)$ and $X_2 = X_3 = V$. We have $\rho_0 = k\frac{|X|}{|X_1|}$, $\rho_1 = k\frac{|X_1|-|X|}{|V|-|X|}$ and $\rho_2 = k$. As before, k is an eigenvalue of $M_4(k; \rho_0, \rho_1, k)$. Since the (i, j) entry of this matrix is null if $i \equiv j \pmod{2}$, the other eigenvalues of $M_4(k; \rho_0, \rho_1, k)$ are $\{\sigma, -\sigma, -k\}$, with $|\sigma| \leq \lambda(G)$. But $\det(M_4(k; \rho_0, \rho_1, k)) = k^2\rho_0(k-\rho_1) = k^2\sigma^2$, and so $\rho_0(k-\rho_1) \leq \lambda(G)^2$. A simple calculation shows that this implies Eq. 1. Note that we implicitly assumed that $N_G(X)$ is a proper subset of V. However, we can directly check that Eq. 1 holds if $N_G(X) = V$.

(3) **Random regular graphs.** In [9], it was shown that, if k is even, then for "most" k-regular graphs G, we have $\lambda_1(G) \leq 2\sqrt{k-1} + O(\log k)$. Therefore, using Theorem 1, we see that for "most" regular graphs, we can prove *in polynomial time* that linear sized subsets have expansion at least

$$\frac{k}{2}\left(1 - \sqrt{1 - \frac{4k-4}{(2\sqrt{k-1} + O(\log k))^2}}\right) = \frac{k}{2} - O(k^{3/4} \log^{1/2} k).$$

(4) **Selection networks.** We can use Theorem 1 to build explicit selection networks of small size. A selection network is a network of comparators that classifies a set of n numbers, where n is even, into two subsets of $n/2$ numbers such that any element in the first set is smaller than any element in the second set. In [18], a probabilistic construction of a selection network is given using an asymptotic upper bound of $2n \log_2 n$ comparators. Also, an upper bound slightly less than $6n \log_2 n$ is shown by a deterministic construction. Using Theorem 1, we can [13] construct selection networks of asymptotic size $(3 + \epsilon)n \log_2 n$, for any $\epsilon > 0$.

(5) **Extrovert graphs.** Given a graph $G = (V, E)$ and a subset X of V, an element of X is said to be extrovert if at least half of its neighbors are outside X. A family of graphs is called extrovert if all linear-sized subsets contain a constant fraction of extrovert nodes. Theorem 2 shows that the average degree of the nodes of a linear-sized induced subgraph of a k-regular Ramanujan graph is at most $(1 + \sqrt{k-1})^+$, which is less than $k/2$ for $k \geq 7$. This shows that Ramanujan graphs of degree at least 7 are extrovert graphs.

6. Acknowledgments

The author is indebted to Tom Leighton for many helpful discussions. He also thanks Noga Alon, Fan Chung, Persi Diaconis, Joel Friedman and Nicholas Pippenger for useful conversations.

References

1. M. Ajtai, J. Komlós, and E. Szemerédi, *Deterministic simulation in logspace*, STOC87.
2. _____, *Sorting in $c \log n$ parallel steps*, Combinatorica **3** (1983), 1–19.
3. N. Alon, *Eigenvalues and expanders*, Combinatorica **6** (1986), no. 2, 83–96.
4. N. Alon and F. R. K. Chung, *Explicit construction of linear sized tolerant networks*, Discrete Mathematics **72** (1988), 15–19.
5. S. Arora, T. Leighton, and B. Maggs, *On-line algorithms for path selection in a non-blocking network*, STOC90.
6. M. Bellare, O. Goldreich, and S. Goldwasser, *Randomness in interactive proofs*, FOCS90.
7. A. Z. Broder, A. M. Frieze, E. Shamir, and E. Upfal, *Near-perfect token distribution*, preprint.
8. F. R. K. Chung, V. Faber, and T. A. Manteuffel, *An upper bound on the diameter of a graph from eigenvalues asociated with its laplacian*, preprint.
9. J. Friedman, *On the second eigenvalue and random walks in random regular graphs*, Combinatorica **11** (1991), no. 4, 331–362.
10. _____, *Some geometric aspects of graphs and their eigenfunctions*, Tech. report, Princeton University, Department of Computer science, 1991.
11. O. Goldreich, R. Impagliazzo, L. Levin, R. Venkatesan, and D. Zuckerman, *Security preserving amplification of hardness*, FOCS90.
12. N. Kahale, *Better expansion for ramanujan graphs*, FOCS91, pp. 398–404.
13. N. Kahale and N. Pippenger, Personal Communication.

14. T. Leighton and B. Maggs, *Fast algorithms for routing around faults on multibutterflies*, FOCS89.
15. A. Lubotzky, R. Phillips, and P. Sarnak, *Ramanujan graphs*, Combinatorica **8** (1988), no. 3, 261–277.
16. G. A. Margulis, *Explicit group-theoretical constructions of combinatorial schemes and their applications to the design of expanders and concentrators*, Problemy Peredači Informacii **24** (1988), no. 1, 51–60.
17. A. Nilli, *On the second eigenvalue of a graph*, Discrete Mathematics **91** (1991), 207–210.
18. N. Pippenger, *Selection networks*, Tech. Report 90-12, The University of British Columbia, Department of Computer science, 1990.
19. P. Sarnak, *Some applications of modular forms*, Cambridge University Press, 1990.
20. E. Seneta, *Non-negative matrices and markov chains*, Springer-Verlag, 1981.
21. R. M. Tanner, *Explicit construction of concentrators from generalized n-gons*, SIAM J. Algebraic Discrete Methods **5** (1984), no. 3, 287–294.
22. E. Upfal, *An $O(\log n)$ deterministic packet routing scheme*, STOC89.
23. L. Valiant, *Graph theoretic properties in computational complexity*, J. Comput. System Sci. **13** (1976), 278–285.

MIT LABORATORY FOR COMPUTER SCIENCE, CAMBRIDGE, MASSACHUSETTS 02139
E-mail address: kahale@theory.lcs.mit.edu

Numerical Investigation of the Spectrum for Certain Families of Cayley Graphs

JOHN LAFFERTY AND DANIEL ROCKMORE

October 15, 1992

ABSTRACT. In this paper we extend some earlier computations [8]. In particular, the expanding behavior of Cayley graphs of $PSL_2(\mathbb{F}_{107})$ is compared with that of the Cayley graphs for the group A_{10}. These computations support the (up to now) unvoiced conjecture of Lubotzky that the symmetric groups and projective linear groups have asymptotically different average expanding behavior. We also give a thorough spectral analysis for a natural family of Cayley graphs which does not admit analysis by Selberg's theorem.

1. Introduction

Spectral analysis and operator theory have provided some of the main tools for the recent advances in constructions of expander graphs. In particular, by exploiting the various relationships between the second largest eigenvalue of the Laplacian and the expansion coefficient of graphs, families of expanders have been constructed and analyzed. When the graphs of interest are Cayley graphs, techniques from Fourier analysis are especially useful in this analysis.

In this paper we extend the analysis of [8] by presenting two different computations. The first computation gives an analysis of the spectrum of the Cayley graphs of $SL_2(\mathbb{F}_p)$ on the generating sets $\left\{\begin{pmatrix} 1 & 2 \\ 0 & 1 \end{pmatrix}, \begin{pmatrix} 1 & 0 \\ 2 & 1 \end{pmatrix}\right\}$ and $\left\{\begin{pmatrix} 1 & 3 \\ 0 & 1 \end{pmatrix}, \begin{pmatrix} 1 & 0 \\ 3 & 1 \end{pmatrix}\right\}$. Viewed as elements of $SL_2(\mathbb{Z})$, the former set generates a subgroup of finite index, while the latter set generates a subgroup of *infinite* index in $SL_2(\mathbb{Z})$. Consequently, for the latter pair, the often useful techniques

---

1991 *Mathematics Subject Classification.* Primary 05C25, 20C40, 68R10 ; Secondary 20B25, 20D06, 20C30.

The second author was supported in part by an NSF Mathematical Sciences Postdoctoral Fellowship.

The final version of this paper will be submitted for publication elsewhere

of Selberg's theorem are of no help in studying the asymptotic spectral behavior of the graphs. Our results in this first computation give further evidence for speculations presented in our earlier work (cf. [8], Section 6).

The second computation compares the average expanding behavior of Cayley graphs of $PSL_2(\mathbb{F}_{107})$ with that of A_{10}. Our motivation for undertaking this computation again comes from our earlier work which served as initial evidence towards the conjecture of Lubotzky that the symmetric groups and special linear groups have asymptotically different average expanding behavior. It is well-known that if a pair of elements of $SL_2(\mathbb{F}_p)$ is chosen at random, then with high probability (approaching 1 as p goes to infinity, at a rate of roughly $1 - O(\log^2 p/p)$) these elements will generate $SL_2(\mathbb{F}_p)$ [6]. In [8] this led us to consider the following question: if two elements a, b are chosen uniformly from $SL_2(\mathbb{F}_p)$ subject to the constraint that $\langle a, b \rangle = SL_2(\mathbb{F}_p)$, then what can be said about $\lambda_1(SL_2(\mathbb{F}_p), \{a, b\})$?[1] Our considerations were from a computational point of view, and letting p vary from 3 to 199 we chose 50 generating pairs at random, and computed λ_1 for each such pair. The results showed a marked clustering away from 1, roughly between 0.865 and 0.890. (For details as well as the graph, see [8], Section 5.3.) The clustering of the these eigenvalues naturally suggests that one should be able to prove that *asymptotically, almost every generating pair of $SL_2(\mathbb{F}_p)$ is a good expander* ([8], Section 6, Question 4).

In a similar vein is an open question concerning the expanding behavior of the symmetric groups. In [9] Lubotzky asks the question, *can the family of symmetric groups be made into a family of expanders for any set of generators?*. Recent work seems to indicate that the answer to this question may be no. So, it was suggested to us by Lubotzky that we consider computing λ_1 for *every* pair of generators for comparably sized symmetric and special linear groups. The possibility of doing this is, of course, dictated by the size of the groups involved. But our experience shows that the spectral behavior of the special linear groups becomes "typical" for p near 100. Our techniques compute the spectrum by computing the appropriate Fourier transforms at a complete set of irreducible representations for $SL_2(\mathbb{F}_p)$ (cf. Section 2). These representations have size on the order of p, so that the eigenvalue computations pose no problems for $p \approx 107$.

For the symmetric groups, our understanding of the asymptotics of numerical spectral analysis is less well-developed. A serious computational consideration for the symmetric groups is that the degrees of the irreducible representations grow quickly [7]. Thus, the determination of the eigenvalues soon dominate the computation. The symmetric group S_{10} is still computationally tractable from this point of view (with largest irreducible representation of degree 768). Its size is comparable with $SL_2(\mathbb{F}_p)$ for $p = 107$. We simplify the problem of computing the spectrum of generating pairs by considering only the 3-regular Cayley graphs for $PSL_2(\mathbb{F}_{107})$ and A_{10}, taking one generating element to be an involution.

---

[1] If $\langle S \rangle = G$, then we write $\lambda_1(G, S)$ for λ_1 of this Cayley graph.

The remainder of this paper is organized as follows. In Section 2 we review the relevance of Fourier analysis for spectral computations of Cayley graphs. In the Section 3 we present a very brief description of the representation theory of the groups $SL_2(\mathbb{F}_p)$, $PSL_2(\mathbb{F}_p)$ and A_n, in order to give an indication of the nature of the computation. In the final sections, we discuss the computations that were carried out. Section 4 describes the analysis of the generating pairs $\left\{ \begin{pmatrix} 1 & 2 \\ 0 & 1 \end{pmatrix}, \begin{pmatrix} 1 & 0 \\ 2 & 1 \end{pmatrix} \right\}$ and $\left\{ \begin{pmatrix} 1 & 3 \\ 0 & 1 \end{pmatrix}, \begin{pmatrix} 1 & 0 \\ 3 & 1 \end{pmatrix} \right\}$, viewed as elements of $SL_2(\mathbb{Z})$. Section 6 describes the distribution of second-largest eigenvalues over all 3-regular Cayley graphs for $PSL_2(\mathbb{F}_{107})$ and A_{10}. The paper concludes by summarizing the results and suggesting directions for further investigation.

We would like to thank Alex Lubotzky for his suggestions and encouragement, as well as Larry Finkelstein for always answering our permutation group questions.

2. Cayley graphs and Fourier analysis

Let G be a finite group and let $S \subset G$ generate G. The *Cayley graph* $X = X(G, S)$ for G with respect to S is the undirected graph with vertex set equal to G, such that there is an edge between a and b in X if and only if $as = b$ for some $s \in S \bigcup S^{-1}$.

The *adjacency matrix* of X is the $|X|$ by $|X|$ matrix, with rows and columns indexed by vertices of X such that the x, y entry is 1 or 0 depending on whether or not x and y are adjacent. The spectrum of a graph is the spectrum of its adjacency matrix.

Assuming that X is k-regular, the largest eigenvalue of X is k. Let $\mu_1 \leq \cdots \leq \mu_{N-1} \leq k$ denote the eigenvalues of X. The *second-largest eigenvalue* of X, denoted $\lambda_1(X)$, is defined to be

$$\lambda_1 = \max_{\{i:\, |\mu_i| \neq k\}} |\mu_i|.$$

Various connectivity properties of a graph can be judged by studying its spectrum (see *e.g.*, [1]). Bounds for the diameter, expansion coefficient, and chromatic number all can be given in terms of λ_1. For details and references we refer the reader to other papers in this volume as well as the books ([9], [11]).

Fourier analysis provides an extremely useful tool for the study of the spectra of Cayley graphs. Recall that if f is any complex-valued function defined on G and ρ is any matrix representation of G, then the Fourier transform of f at ρ is defined to be the matrix sum

$$\widehat{f}(\rho) = \sum_{t \in G} f(t)\rho(t).$$

Let δ_S denote the characteristic function of the set $S \bigcup S^{-1}$. Then it is immediate that the adjacency matrix of $X(G, S)$ is equal to $\widehat{\delta_S}(\rho_{\text{reg}})$, up to a reordering of the group elements, where ρ_{reg} denotes the right regular representation of G.

Block diagonalization of the regular representation may be obtained using representation theory. If $\{\rho_1, \ldots, \rho_h\}$ is a complete set of inequivalent irreducible matrix representations of G, then there exists a change of basis such that

$$\widehat{\delta_S}(\rho_{\text{reg}}) \sim \begin{pmatrix} B_1 & 0 & \cdots & 0 \\ 0 & B_2 & \cdots & 0 \\ \vdots & \vdots & \ddots & \vdots \\ 0 & 0 & \cdots & B_h \end{pmatrix}$$

where

$$B_i = \begin{pmatrix} \widehat{\delta_S}(\rho_i) & 0 & \cdots & 0 \\ 0 & \widehat{\delta_S}(\rho_i) & \cdots & 0 \\ \vdots & \vdots & \ddots & \vdots \\ 0 & 0 & \cdots & \widehat{\delta_S}(\rho_i) \end{pmatrix}$$

with $\deg(\rho_i)$ copies of $\widehat{\delta_S}(\rho_i)$ on the diagonal. Consequently,

$$(2.1) \qquad \text{spectrum}(X(G, S)) = \bigcup_{i=1}^{h} \text{spectrum}(\widehat{\delta_S}(\rho_i)).$$

For numerical analysis of the spectrum, (2.1) provides a great advantage. Standard techniques for computing eigenvalues of a square matrix of degree d require $O(d^3)$ operations [13]. As the degree of the largest irreducible representation is bounded by $|G|^{1/2}$, (2.1) brings the computation from $O(|G|^3)$ operations to a more manageable $O(|G| \cdot (\max_i \deg(\rho_i)))$ operations.

In the case in which $S \bigcup S^{-1}$ is a union of conjugacy of classes of G, $\widehat{\delta_S}$ can be completely diagonalized and the diagonal elements computed as certain character sums [3]. However, if $S \bigcup S^{-1}$ is not a union of conjugacy classes (which is the situation in the computations considered here), then the actual matrix representations are required to compute the spectrum.

3. Representation theory for $SL_2(\mathbb{F}_p)$, $PSL_2(\mathbb{F}_p)$ and A_n.

In this section we give a very quick introduction and explanation of the representation theory of $SL_2(\mathbb{F}_p)$, $PSL_2(\mathbb{F}_p)$ and A_n. In-depth treatments can be found in [4] for A_n and [10] for $SL_2(\mathbb{F}_p)$ and $PSL_2(\mathbb{F}_p)$.

3.1. Representation theory for the alternating group.
The representation theory of the alternating groups is most easily explained by using the representation theory of the symmetric groups.

A *partition* $\lambda \vdash n$ of a natural number n is a tuple $\lambda = (\lambda_1, \lambda_2, \ldots, \lambda_h)$ with $\lambda_i \geq \lambda_{i+1} > 0$ and $\sum_i \lambda_i = n$. The irreducible representations of the symmetric group S_n are parametrized by the partitions of n. For any such λ there is a corresponding *Ferrer's Diagram*, which is a left-justified arrangement of rows of boxes with λ_i boxes in row i. Figure 1 shows the Ferrer's diagram for the partition $(4, 2, 2, 1) \vdash 9$.

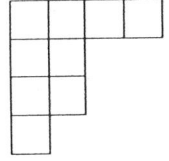

FIGURE 1: The Ferrer's diagram for $(4, 2, 2, 1)$

If $\lambda \vdash n$, then a *Young tableau* of shape λ is a Ferrer's diagram for λ with the numbers $1, \ldots, n$ inserted in the boxes. Two tableaux are said to be *row-equivalent* if any two corresponding rows contain the same sets of entries. A *tabloid* is an equivalence class of tableaux. S_n acts naturally on the set of all tabloids of a given shape, permuting the entries. The associated permutation module of S_n is denoted by M^λ. M^λ is reducible, except for the trivial partition (n). There is a natural way in which to pick out a specific irreducible component of M^λ, called the *Specht module*, and denoted S^λ. It is the action of S_n on S^λ which gives the irreducible representation ρ_λ associated to λ. For details see [4].

By restriction, any representation of S_n, gives a representation of A_n. It turns out that most of the ρ_λ remain irreducible upon restriction.

If $\lambda \vdash n$, then the *dual* partition, denoted λ' is the partition associated to the tranpose of the diagram for λ. For example, $(4, 2, 2, 1)' = (4, 3, 1, 1)$.

THEOREM 3.1. *Let $\lambda, \nu \vdash n$. Then $\rho_\lambda \downarrow A_n = \rho_\nu \downarrow A_n$ iff either $\lambda = \nu$ or $\lambda' = \nu$. If $\lambda' \neq \lambda$ then $\rho_\lambda \downarrow A_n$ remains irreducible. Otherwise, $\rho_\lambda \downarrow A_n$ splits into two inequivalent irreducible components. This gives all the irreducible representations of A_n.*

Construction of the irreducible representations is straightforward, and orthogonal matrices (Young's orthogonal form) for the pairwise adjacent transpositions can be written down in terms of certain parameters from the tabloids ([4], Section 3.2). In fact for these elements, the matrices are even sparse, greatly expediting computations.

3.2. Representation theory for $SL_2(\mathbb{F}_p)$ and $PSL_2(\mathbb{F}_p)$. The irreducible representations of $SL_2(\mathbb{F}_p)$ occur in two families, the *discrete series* and *principal series*. The distinction depends upon the restriction of an irreducible representation to the Borel subgroup $B < SL_2(\mathbb{F}_p)$ of upper triangular matrices. An irreducible representation of $SL_2(\mathbb{F}_p)$ is said to be from the principal series if its restriction to B contains the trivial representation. Otherwise, it is said to be from the discrete series.

Knowledge of the representations of $SL_2(\mathbb{F}_p)$ gives the irreducible representations of $PSL_2(\mathbb{F}_p)$.

THEOREM 3.2. *Let ρ be an irreducible matrix representation of $SL_2(\mathbb{F}_p)$ and let $I = \begin{pmatrix} 1 & 0 \\ 0 & 1 \end{pmatrix} \in SL_2(\mathbb{F}_p)$. If $-I$ is in the kernel of ρ, then ρ is constant on cosets $SL_2(\mathbb{F}_p)/\{\pm I\}$ and as such gives an irreducible representation of $PSL_2(\mathbb{F}_p)$. Under this identification, the set $\{\rho \mid \{\pm I\} \subset ker(\rho)\}$ gives a complete set of inequivalent irreducible representations of $PSL_2(\mathbb{F}_p)$.*

As for the symmetric group, the irreducible representations for $SL_2(\mathbb{F}_p)$ may be effciently constructed. Explicit contructions are given in [8], using the discussions in [10] and [12]. In short, all representations of the principal series can be constructed as induced 1-dimensional representations from B and as such are essentially in 1-1 correspondence with the characters of the diagonal subgroup, or split torus. The discrete series is less easily explained, but suffice it to say here that the representations are in close correspondence to the characters of the non-split torus in $SL_2(\mathbb{F}_p)$ (cf. [10], Ch. 2, Section 5). From a computational point of view, the discrete series representations are much more expensive to construct. However, our experience has shown that the spectral properties of the principal and discrete series representations are almost identical. Consequently, in the computations discussed in Section 5 for computing all 3-regular Cayley graphs for $PSL_2(\mathbb{F}_p)$, we actually carried out the computation by evaluating the Fourier transforms only at principal series representations.

4. Analysis of two infinite families of Cayley graphs

In this section we consider the spectrum for the sets of generators (of $SL_2(\mathbb{F}_p)$)

$$\mathcal{G}_2 = \left\{ \begin{pmatrix} 1 & 2 \\ 0 & 1 \end{pmatrix}, \begin{pmatrix} 1 & 0 \\ 2 & 1 \end{pmatrix} \right\}, \text{ and } \mathcal{G}_3 = \left\{ \begin{pmatrix} 1 & 3 \\ 0 & 1 \end{pmatrix}, \begin{pmatrix} 1 & 0 \\ 3 & 1 \end{pmatrix} \right\}.$$

This extends our earlier work ([8], Section 5.2) and interestingly, shows that the open questions and speculations presented there also obtain for families which do not seem to admit analysis by Selberg's Theorem; that is, families which do not come from subgroups of finite index in $SL_2(\mathbb{Z})$.

Figure 2 plots $\lambda_1(\mathcal{G}_i, SL_2(\mathbb{F}_p))$ versus p for $i = 2, 3$. The solid line at $\sqrt{3}/2$ marks the Ramanujan bound for these 4-regular graphs.

Figure 3 displays the full spectrum of $X(\mathcal{G}_i, SL_2(\mathbb{F}_p))$ ($i = 2, 3$) as a function of p. The graphs on the right display all eigenvalues occurring in discrete series representations of $SL_2(\mathbb{F}_p)$ (with $\mathcal{G}_2$ on top and $\mathcal{G}_3$ on the bottom), while the graphs on the left show all which occur in principal series representations. That is, for every prime p, a single "dot" is displayed at the point (λ, p) if λ occurs as an eigenvalue for $X(\mathcal{G}_i, SL_2(\mathbb{F}_p))$ in either principal series Fourier transform (graphs on the left of Figure 3) or a discrete series Fourier transform (graphs on the right of Figure 3). Note the remarkable similarity in the graphs. Even more

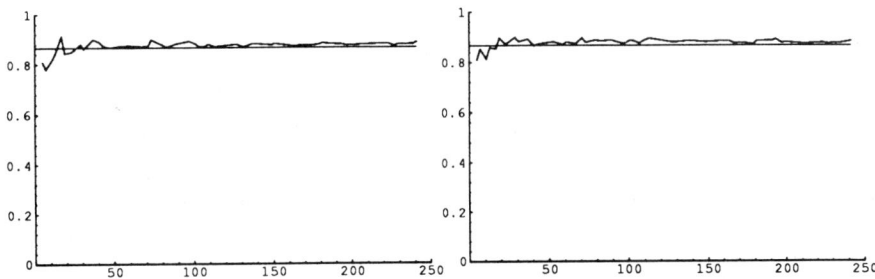

FIGURE 2: $\lambda_1(\mathcal{G}_2, SL_2(\mathbb{F}_p))$ (left), $\lambda_1(\mathcal{G}_3, SL_2(\mathbb{F}_p))$ (right)

than this appears to be true, as computation shows that if only the eigenvalues in a single Fourier transform are considered, then they are in fact generic, in the sense that if displayed as in Figure 3, they would yield similar pictures. See [8], Section 6, Question 2. Computational considerations for generating these graphs may be found in [8], Sections 3 and 4.

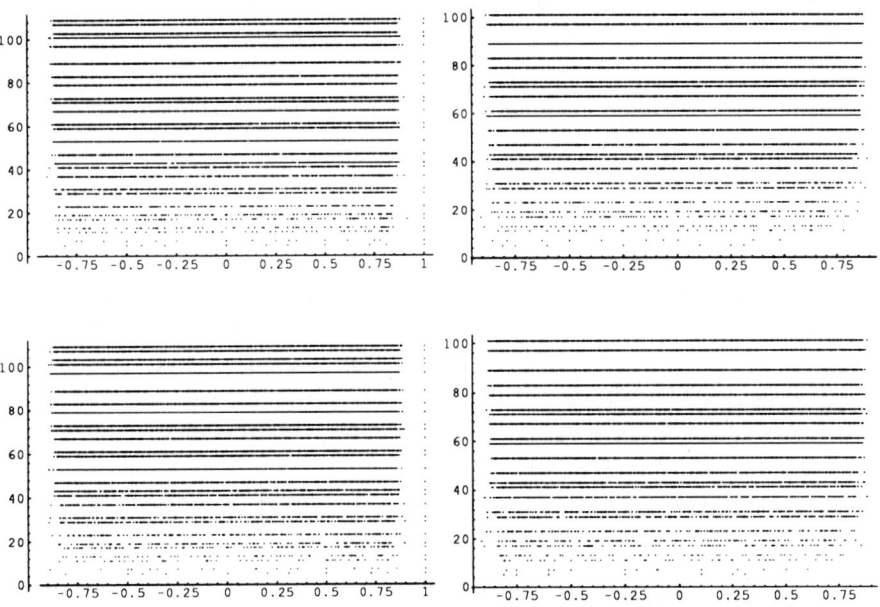

FIGURE 3: Principal series spectrum (left) and discrete series spectrum (right) for $\lambda_1(\mathcal{G}_2, SL_2(\mathbb{F}_p))$ (top) and $\lambda_1(\mathcal{G}_3, SL_2(\mathbb{F}_p))$ (bottom).

5. Distribution of λ_1 for A_{10} and $PSL_2(\mathbb{F}_{107})$

In this section we compare the distribution of λ_1 for all 3-regular Cayley graphs for the group A_{10} with that of all 3-regular Cayley graphs for the comparably sized $PSL_2(\mathbb{F}_{107})$. The motivation is to quantify the extent to which random Cayley graphs for the groups $PSL_2(\mathbb{F}_p)$ and A_n are good expanders. While it is known by elementary combinatorial arguments that general random graphs are expanders, little is known about the behavior for Cayley graphs of the classical finite groups. The computation described in this section naturally extends the investigations of [8], where it was seen that 4-regular graphs for $SL_2(\mathbb{F}_p)$ tend to have expansion coefficient close to the Ramanujan bound for p larger than 100. In this section we present the complete distribution of λ_1 for 3-regular Cayley graphs for $PSL_2(\mathbb{F}_{107})$ and compare it to the corresponding distribution for A_{10}.

5.1. Determining generating pairs. For our analysis of all 3-regular Cayley graphs (cf. Section 3.2) we need an efficient test of whether or not, for a given $\{a,b\} \in G$, we have $\langle a,b \rangle = G$. In case $G = PSL_2(\mathbb{F}_p)$ we use the classification of subgroups of $SL_2(\mathbb{F}_p)$. In brief, this is done by checking the orbit structure of the action of the generators (lifted to $SL_2(\mathbb{F}_p)$) on the projective line, where the action should be transitive. This is explained fully in [8].

For $G = A_n$, we use the fact [5] that if a permutation group on $n > 5$ points is transitive on 4-sets (4-homogeneous), then outside of a few exceptions, the group is either A_n or S_n. The orbit structure of the action of the group generated by two elements on the collection of 4-sets is easily determined on the computer.

To simplify the computation we take advantage of the fact that the spectrum is invariant under conjugation. If $S \subset G$, and $t \in G$ then let ${}^tS = tSt^{-1}$. Notice that $\lambda_1(G,S) = \lambda_1(G, {}^tS)$, since

$$\sum_{s' \in {}^tS} \rho_{\text{reg}}(s') = \rho_{\text{reg}}(t) \left(\sum_{s \in S} \rho_{\text{reg}}(s) \right) \rho_{\text{reg}}(t^{-1}).$$

Consequently, if $\alpha_1, \ldots, \alpha_k$ denote a set of conjugacy class representatives for involutions in G then at the very worst we need compute for all pairs of the form $\{\alpha_i, b\}$ for $b \in G$. A further reduction is obtained by then using the fact that for any $t \in Z_G(\alpha_i)$ (the centralizer of α_i), ${}^t\{\alpha_i, b\} = \{\alpha_i, {}^tb\}$. To implement these observations, for each involution α_i we decompose G into equivalence classes of orbits of $Z_G(\alpha_i)$. This can be accomplished in $O(|G|)$ operations. Then, we chose a representative b of each each orbit and compute the spectrum for the pair $\{\alpha_i, b\}$, computing a total of $|G| / |Z_G(\alpha_i)|$ eigenvalues.

5.2. The cumulative distribution of λ_1. Figure 4 presents the graphs of the cumulative distribution functions $F_G(\lambda)$ for $G = A_{10}$ and $G = PSL_2(\mathbb{F}_{107})$ where

$F_G(\lambda) =$ the fraction of generating pairs $S = \{a,b\}$ such that $\langle a,b \rangle = G$ and $a^2 = 1$ with $\lambda_1(G, \{a,b\}) \leq \lambda$.

In our current implementation, working with representations of the discrete series is more computationally intensive by a factor of p than working with those of the principal series. Thus, the eigenvalues were computed only from the principal series. This approximation was justified by our experience [8] that the principal and discrete series representations are essentially identical from the point of view of spectral analysis of Cayley graphs.

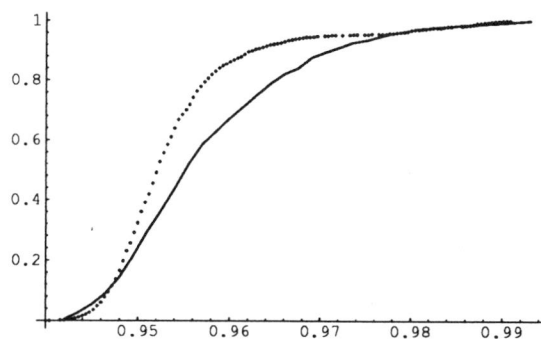

FIGURE 4: $F_\lambda(PSL_2(\mathbb{F}_{107}))$ (dashed); $F_\lambda(A_{10})$ (solid)

The figure shows that the eigenvalues for $PSL_2(\mathbb{F}_{107})$ are more sharply concentrated near the Ramanujan bound of $\frac{2\sqrt{2}}{3} \approx 0.942809$ than those for A_{10}. As Figure 8 of [8] indicates, this concentration becomes much greater for $PSL_2(\mathbb{F}_p)$ as p grows larger. The figure suggests that the distribution of second-largest eigenvalues is qualitatively different for the two groups. However, the computation is, of course, inconclusive, and further investigation of the asymptotics will be required to resolve the question.

As discussed above, the computation was carried out by choosing for each involution, a representative of each orbit of the associated centralizer. Since $PSL_2(\mathbb{F}_{107})$ has only one conjugacy class of involutions, while A_{10} has two, it is conceivable that the two classes in A_{10} may behave differently with respect to expansion. We examined this possibility by comparing the respective spectra. Figure 5 plots λ_1 for each of these conjugacy classes. It is seen that the distribution of eigenvalues associated with the involution $(12)(34)$ is characteristic of the combined distribution. This is to be expected, however, due to the fact that the number of conjugacy classes of generating pairs is 1496 for this involution, while there are only 82 such classes for the involution $(12)(34)(56)(78)$.

Finally, it has been observed that the action of $SL_2(\mathbb{F}_p)$ on the projective line approximates the action of $SL_2(\mathbb{F}_p)$ on itself ([2], [8]). That is, $SL_2(\mathbb{F}_p)$ acts naturally on $\mathbb{P}^1(\mathbb{F}_p)$. Let $\langle a, b \rangle = SL_2(\mathbb{F}_p)$ and consider the graph with vertex set $\mathbb{P}^1(\mathbb{F}_p)$ and $v \sim w$ iff $sv = w$ for some $s \in \{a, b\} \cup \{a, b\}^{-1}$. The conjecture is that λ_1 for this graph closely approximates $\lambda_1(G, \{a, b\})$. This is equivalent to saying that the largest nontrivial eigenvalue occurring in the Fourier transform at $1 \uparrow_B^{SL_2(\mathbb{F}_p)}$ is close to $\lambda_1(G, \{a, b\})$. The analogous case for symmetric groups would be that natural permutation representation of A_n acting on $\{1, \ldots, n\}$ approximates the action of the full set of irreducible representations. Figure 6 seems to show that this is not the case for A_{10}.

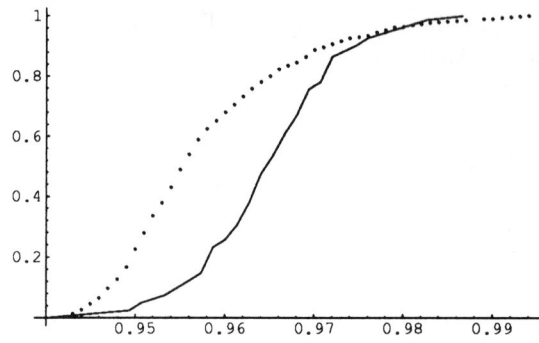

FIGURE 5: λ_1 for 3-regular Cayley graphs on A_{10},
dashed–involution (12)(34); solid–involution (12)(34)(56)(78)

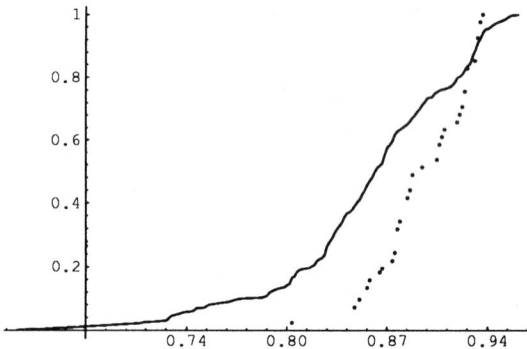

FIGURE 6: λ_1 for the action on the permutation graph for A_{10},
dashed–involution (12)(34); solid–involution (12)(34)(56)(78)

6. Conclusions and open questions

The computations that have been described in this paper cast a faint light on the spectral geometry of finite groups. Clearly, further theorems are needed to illuminate the phenomena that we see in the data. In this section we suggest several areas for further investigation.

Our computations suggest that there is a significant difference in the asymptotic behavior of the spectra of Cayley graphs for $PSL_2(\mathbb{F}_p)$ and A_n. This difference should be quantified. For example, it is natural to ask whether a lower bound of the form

$$\liminf_{n,p\to\infty} \| F_\lambda(PSL_2(\mathbb{F}_p)) - F_\lambda(A_n) \|_1 > c$$

can be obtained, where the L^1 distance could be replace by any other suitable measure of similarity, such as the Kullback-Liebler distance.

In this paper we have restricted our attention to 3-regular Cayley graphs, where one of the generating elements is an involution. It is natural to ask if a general k-regular Cayley graph enjoys the same properties that appear to hold in the cases we have considered. In particular, it would be of interest to determine whether or not a k-regular Cayley graph for $PSL_2(\mathbb{F}_p)$ is a good expander with high probability as $p \to \infty$.

In addition, we have observed [8] that the action of $PSL_2(\mathbb{F}_p)$ on itself is closely approximated by its action on the projective line $\mathbb{P}^1(\mathbb{F}_p)$, but an analogous statement is perhaps not true for A_n. Thus, the question arises: for which classical finite groups G is there a finite set S on which G acts, with $\frac{|S|}{|G|} = o(1)$ as $|G| \to \infty$, such that the spectrum of $X(G, \{a, b\})$ is closely approximated by the graph of the action of $\{a, b\}$ on S? Since the computational payoff from having such an approximation is so great, an understanding of this question would be of considerable interest.

Finally, our computations show that the generating sets $\mathcal{G}_2$ and $\mathcal{G}_3$ for $SL_2(\mathbb{F}_p)$ result in Cayley graphs having identical spectral properties, in spite of the fact that Selberg's theorem only applies to the first set. Thus, we are led to inquire to what extent the property of being a quotient of a subgroup of finite index in $SL_2(\mathbb{Z})$ determines expansion properties. This example would also seem to indicate that there should be a proof of the expansion properties for such families that does not depend on Selberg's theorem.

References

1. N. Biggs *Algebraic Graph Theory*, Cambridge Univ. Press, Cambridge, 1974.
2. M. Buck, *Expanders and diffusers.*, SIAM J. Alg. and Disc. Methods **7** (1986), 282-304.
3. P. Diaconis, *Group Representations in Probability and Statistics*, Inst. of Math. Stat., Hayward, CA, 1989.
4. G. D. James and A. Kerber, *The Representation Theory of the Symmetric Group*, Ency. of Math. and its Applications, Vol. 16, Addison-Wesley, Reading, Mass., 1981.
5. W. Kantor, *4-homogeneous groups*, Math. Zeit., **103** (1968), 67-68.
6. W. Kantor and A. Lubotzky, *The probability of generating a finite classical group*, Geom. Ded., **36** (1990), 67–87.
7. S. V. Kerov and A. M. Vershik, *Asymptotics of the largest and typical dimensions of the irreducible representations of a symmetric group*, Func. An. and App. **19** (1985), 21–31.
8. J. Lafferty and D. Rockmore, *Fast Fourier analysis for SL_2 over a finite field and related numerical experiments*, Experimental Mathematics **1** (1992), 115–139.
9. A. Lubotzky, *Discrete Groups, Expanding Graphs, and Invariant Measures*. To appear.
10. M. Naimark and A. Stern, *Theory of Group Representations*, Springer-Verlag, NY, 1982.
11. P. Sarnak, *Some Applications of Modular Forms*, Cambridge Univ. Press, Cambridge, 1990.
12. A. Silberger, *An elementary construction of the representations of $SL(2, GF(q))$*, Osaka J. Math. **6** (1969), 329-338.
13. J. Wilkinson, *The Algebraic Eigenvalue Problem*, Oxford Univ. Press, 1965.

IBM, T.J. Watson Research Center, Yorktown Heights, NY 10598

Department of Mathematics and Computer Science, Dartmouth College, Hanover, New Hampshire 03755

Some Algebraic Constructions of Dense Graphs of Large Girth and of Large Size

FELIX LAZEBNIK AND VASILIY A. USTIMENKO

ABSTRACT. For any prime power $q \geq 3$, we consider two infinite series of bipartite q-regular edge-transitive graphs of orders $2q^3$ and $2q^5$ which are induced subgraphs of regular generalized 4-gon and 6-gon, respectively. We compare these two series with two families of graphs, $H_3(p)$ and $H_5(p)$, p is a prime, constructed recently by Wenger ([26]), which are new examples of extremal graphs without 6- and 10-cycles respectively. We prove that the first series contains the family $H_3(p)$ for $q = p \geq 3$. Then we show that no member of the second family $H_5(p)$ is a subgraph of a generalized 6-gon. Then, for infinitely many values of q, we construct a new infinite series of bipartite q-regular edge-transitive graphs of order $2q^5$ and girth 10. Finally, for any prime power $q \geq 3$, we construct a new infinite series of bipartite q-regular edge-transitive graphs of order $2q^9$ and girth $g \geq 14$. Our constructions were motivated by some results on embeddings of Chevalley group geometries in the corresponding Lie algebras and a construction of a blow-up for an incident system and a graph.

Introduction

The missing definitions of graph-theoretical concepts which appear in this paper can be found in [6]. All graphs we consider are simple, i.e. undirected without loops and multiple edges. Let $V(G)$ and $E(G)$ denote the set of vertices and the set of edges of G, respectively. $|V(G)|$ is called the *order* of G, and $|E(G)|$ is called the *size* of G. A path in G is called *simple* if all its vertices are distinct. When it is convenient, we shall identify G with the corresponding antireflexive symmetric binary relation on $V(G)$, i.e. $E(G) \subset V(G) \times V(G)$. The *length* of a path is the number of its edges. The group of all automorphisms of graph G will be denoted by $Aut(G)$. The *girth* of a graph G, denoted by $g = g(G)$, is the length of the shortest cycle in G.

1991 *Mathematics Subject Classification*. Primary 05C35, 05C38; Secondary 05B25.
This research was partially supported by a grant DMS-9020485.
The final version of this paper will be submitted for publication elsewhere

© 1993 American Mathematical Society
1052-1798/93 $1.00 + $.25 per page

Examples of graphs with large girth which satisfy certain additional conditions are known to be hard to construct, and they turn out to be useful in various problems in extremal graph theory, in studies of graphs with high degree of symmetry, and in designs of communication networks. There are many references on each of these topics. Here we mention just a few main books and survey papers which also contain extensive bibliographies. On the extremal graph theory: [6,7,14,29]; on graphs with high degree of symmetry: [9,15,17,27,30,31,32,39]; on communication networks: [2,12].

Let $\mathcal{F}$ be a family of graphs. By $ex(v, \mathcal{F})$ we denote the greatest number of edges in a graph on v vertices which contains no subgraph isomorphic to a graph from $\mathcal{F}$. Let C_m denote the cycle of length $m \geq 3$. According to a well known unpublished result of Erdös (The Even Circuit Theorem), see [29], $ex(v, C_{2k}) = O(v^{1+1/k})$ (for a generalization of this result see [7,14]). This upper bound is known to be sharp for C_4, C_6 and C_{10}. The corresponding construction for C_4 can be found in [10,13,29]. The constructions for C_6 and C_{10} (see [1,29]) are incidence graphs for generalized n-gons, $n = 4, 6$ (geometries of the Chevalley groups $B_2(q)$ and $G_2(q)$). Recently new important examples of graphs with no 6– or 10–cycles were found by Wenger in [40], where they are denoted by $H_3(p)$ and $H_5(p)$ respectively, p is a prime number. These graphs are members of a family $\{H_i(p), i \geq 1\}$, of regular bipartite graphs whose vertex sets are disjoint unions of two i-dimensional vector spaces over the prime field $\mathbb{F}_p$, and whose edges are defined by certain systems of equations.

The content of this paper is outlined below.

(i) In Section 1 we present a construction of a blow–up of a graph which is used in subsequent sections.

(ii) In Section 2 we consider a connection between Wenger graphs and generalized n-gons. Let I be the incidence relation and $\{p, l\}$ be a flag of the regular generalized n–gon, $n = 4, 6$, over $\mathbb{F}_q$, $q = p^m$. We consider graphs $S_n(q)$ obtained by restricting I on the set $P_n \cup L_n$, where P_n (L_n) is the set of points (lines) opposite to p (l) in the n-gon. Coordinatizations of the generalized n–gons (see [33], [34]) allow to identify each P_n and L_n with a vector space $\mathbb{F}_q^{n-1}$, and the incidence of vectors from P_n and L_n can be expressed in terms of systems of equations on their coordinates. If $n = 4$, this system coincides with the one for $H_3(p)$. Therefore $S_4(q)$ is a simple generalization of $H_3(p)$. On the other hand, if $n = 6$ and $q = p > 2$, graphs $S_6(p)$ and $H_5(p)$ are not isomorphic: graph $H_5(p)$ contains an 8–cycle, hence it cannot be isomorphic to a subgraph of a generalized 6–gon.

(iii) In Section 3, we construct an infinite series of regular bipartite edge–transitive graphs of girth 10. Having girth 10, they cannot be isomorphic to subgraphs of the generalized 6–gons, but they have asymptotically as many edges as regular generalized 6–gons.

It is known that $ex(v, \{C_3, C_4, \ldots, C_m\}) \geq c_m v^{1+\frac{1}{m-1}}$ for some positive constant c_m, $m \geq 3$. This result follows from a theorem proved implicitly by Erdös (see [29]) and the proof is nonconstructive. As it was mentioned in [29], it is unlikely that this lower bound is sharp. For any prime power $q \geq 3$, we construct a q-regular bipartite graph $G(q)$ of order $v = 2q^9$, size $e = q^{10}$ and girth ≥ 14, which supports this claim. For these graphs $e \sim 2^{\frac{10}{9}} v^{1+\frac{1}{9}}$, which is better than the best previously known lower bound $c_{13} v^{1+\frac{1}{12}}$. (See also Section 4 and [37,41]). Graph $G(q)$ is also edge–transitive.

(iv) Finally, in Section 4, we generalize the construction for $G(q)$, and build a new infinite series of regular bipartite graphs with edge-transitive automorphism group and large girth. More precisely, for any positive odd integer $k \geq 3$ and any prime power q, we build a q-regular bipartite graph $D(k, q)$ on $2q^k$ vertices with girth $g \geq k + 5$. This series is an example of a "series of graphs with large girth", and to our knowledge, for $k \geq 19$, it is "the second best" known explicit example of such a series. More details are given in Section 4.

Our constructions were motivated by some results on embeddings of Chevalley group geometries in the corresponding Lie algebras [34,35], and a construction of a blow–up for an incidence system and a graph [33,36].

1. A blow–up of the graph

For a positive integer $n \geq 1$, let $[n] = \{1, 2, \ldots, n\}$ and $2^{[n]}$ denote the set of all subsets of $[n]$. Let $\mathcal{L}$ be an n–dimensional vector space over some field K with a fixed basis $\{e_i \mid i \in [n]\}$. For an arbitrary subset A of $[n]$, let $\mathcal{L}_A$ denote the subspace of $\mathcal{L}$ spanned by $\{e_i \mid i \in A\}$. By $x|_A$ we denote the canonical projection of a vector $x \in \mathcal{L}$ on $\mathcal{L}_A$. Let G be a graph, and let $\eta : V(G) \to 2^{[n]}$ be a mapping of the set of vertices of G into $2^{[n]}$. Finally, let $*$ denote a skew–symmetric bilinear product on $\mathcal{L}$. Consider a new graph $\widetilde{G}$ with the vertex set $\widetilde{V}$ defined as

$$\widetilde{V} = \{(a, x) | a \in V(G), x \in \mathcal{L}_{\eta(a)}\}.$$

We define two distinct vertices (a, x) and (b, y) of $\widetilde{G}$ to be adjacent if and only if

$$\{a, b\} \in E(G) \quad \text{or} \quad x^{h_b} - y^{h_a}|_{\eta(a) \cap \eta(b)} = x * y|_{\eta(a) \cap \eta(b)},$$

where, for $a \in V(G)$, $h_a : e_i \mapsto \lambda_a(i) e_i$, $i \in [n]$, is a nonsingular diagonal operator of $\mathcal{L}$ (defined by its action on the vectors from the basis). We call graph $\widetilde{G}$ a *blow–up* of G.

In [33,34], Ustimenko showed that the incidence relation of the geometry $\gamma(G)$ of a Chevalley group G is a blow–up of the geometry $\gamma(W)$ of its Weyl group W. In this case the vector space $(\mathcal{L}, *)$ is the Lie algebra $\mathcal{L}^+ = \sum_{\alpha \in \Phi^+} \mathcal{L}_\alpha$, where Φ^+ is the set of positive roots for W and $\mathcal{L}_\alpha$ is a root subalgebra of $\mathcal{L}$. The basis vectors $e_\alpha, \alpha \in \Phi^+$, are elements of the Chevalley basis. In particular, each regular generalized n–gon, $n = 3, 4, 6$, corresponding to a Chevalley group of rank 2 of normal type is a blow–up of the incidence graph of the ordinary n–gon (geometry of the dihedral group D_{2n}), which is the cycle C_{2n}. This illustrates that by "blowing up" (over $\mathbb{F}_q$) a small bipartite graph one can obtain a graph of high girth and of large size.

All graphs in this paper are blow–ups of $K_{1,1}$ over finite fields, where $K_{1,1}$ is a graph with two vertices joined by an edge. The bilinear product on $\mathcal{L}$ is defined on the basis elements as $e_i * e_j = \lambda_k \cdot e_k$, where k depends on i and j. For every point p and line l, $|\eta(p)| = |\eta(l)| = n - 1$. Let us assume that $\eta(p) = [n] \setminus \{2\}$ and $\eta(l) = [n] \setminus \{1\}$. Then the set of vertices of the bipartite graph $\widetilde{K}_{1,1}$ can be thought a disjoint union of sets P (set of points) and L (set of lines) of the form $P = \{(x_1, x_3, \ldots, x_n) \mid x_i \in \mathbb{F}_q\}$, $L = \{[y_2, y_3, \ldots, y_n)] \mid y_i \in \mathbb{F}_q\}$.

All graphs in this paper have a group theoretical interpretation as follows. For every $i \in [n]$ and $x \in \mathbb{F}_q$, there exists an automorphism $t_i(x)$ of the graph which acts on the coordinates of vectors of $P \cup L$ by the rule: $x_j \to P_i^j(x, x_1, x_3, \ldots, x_n)$, $y_j \to L_i^j(y, y_2, y_3, \ldots, y_n)$, where P_i^j and L_i^j are polynomials over $\mathbb{F}_q$. The automorphisms $t_i(x)$ satisfy the following properties:

(a) $t_i(x) \cdot t_i(y) = t_i(x + y)$, and so they are the "generalized exponents", and the group $U_i = \langle t_i(x) \mid x \in \mathbb{F}_q \rangle$ is isomorphic to the additive group of $\mathbb{F}_q$.

(b) Group U generated by all $t_i(x)$ is nilpotent and of order q^{n+1} ("generalized unipotent subgroup")

(c) Graph $\widetilde{K}_{1,1}$ is isomorphic to the incidence graph of the following incident structure: sets P and L are the sets of cosets of U with respect to subgroups U_1 and U_2 respectively, with two cosets (one from P, another from L) being incident if and only if their intersection is nonempty.

2. Extremal regular induced subgraphs of generalized 4–and 6–gons

The *incidence structure* (P, L, I) is a triple where P and L are two disjoint sets (set of *points* and set of *lines*, respectively), and I is a symmetric binary relation on $P \cup L$ (*incidence relation*). As is usually done, we impose the following restrictions on I: two points (lines) are incident if and only if they coincide. Let $B = B((P, L, I))$ be a bipartite graph such that $V(B) = P \cup L$ and $E(B) = \{\{p, l\} \mid pIl, p \in P, l \in L\}$. We notice that, according to our definition, B is a simple bipartite graph. We call B the *incidence* graph for the incidence structure (P, L, I).

Let P and L be the sets of vertices and sides of an ordinary n–gon, and I be the natural relation of incidence of a vertex and a side. It is easy to see that the incidence graph of this incidence structure is the cycle C_{2n}. Tits [31] introduced the following definition of a *generalized n-gon* as an incidence structure satisfying the following properties:

(i) for any two distinct elements a and b from $P \cup L$ there exists a positive integer s, $s \leq n$, and a sequence $x_0, x_1, \ldots, x_s$ of distinct elements of $P \cup L$ where $x_0 = a$, $x_s = b$, and $x_i I x_{i+1}$ for $i = 0, \ldots, s - 1$.

(ii) if $s < n$, then the sequence defined in (i) is unique.

Of course, the ordinary ("geometrical") n–gon is a generalized n–gon, and the girth of the incidence graph of a generalized n–gon is $2n$. It is known ([15]), that apart from the ordinary polygons, finite generalized n–gons exist only for $n = 3, 4, 6, 8, 12$.

Some other examples of generalized n–gons for $n = 3, 4, 6$ are closely connected to Chevalley groups $A_2(q), B_2(q), G_2(q)$ of rank 2 over the finite field $\mathbb{F}_q$ (see [11]).

Let G be a Chevalley group of rank 2 over the field $\mathbb{F}_q$, $q = p^m$, p is prime, $m \geq 1$. Then a Borel subgroup of G is the normalizer in G of a Sylow p–subgroup of G. There are exactly two maximal subgroups P_1 and P_2 of G which contain a fixed Borel subgroup B (see [11]). Let us consider the incidence structure (P, L, I), where P is $(G : P_1)$ – the totality of all left cosets of G by P_1, L is $(G : P_2)$, and elements a and b of $P \cup L$ are incident if and only if the intersection of a and b as cosets of G is nonempty. It can be shown, e.g. see [32], that this incidence structure is a generalized n–gon. The corresponding bipartite incidence graph, which we denote by $B_n(q)$, is $(q+1)$–regular.

Let us consider the orbits of the Borel subgroup B on the sets P and L for our generalized n–gons. The cardinalities of orbits on the set of points and the set of lines are the same and equal $1, q, q^2, \ldots q^{n-1}$ (see [11]). Let $S(P)$ and $S(L)$ be the orbits of largest size q^{n-1} on P and L respectively, and $S_n(q)$ be the subgraph of $B_n(q)$ induced on the set $S(P) \cup S(L)$. The importance of graphs $S_n(q)$ in extremal graph theory stems from the fact that they are of girth $2n$ and of size $O(v^{1+\frac{1}{n-1}})$.

Theorem 2.1. *For $n = 4, 6$, graph $S_n(q)$, satisfies the following properties:*
(a) $S_n(q)$ is q–regular of order $2q^{n-1}$ and size q^n
(b) $S_n(q)$ is a graph of girth $2n$.
(c) $S_n(q)$ is edge-transitive
(d) For $q = 2^k, k \geq 1$, $S_4(q)$ is vertex-transitive. For $q = 3^k, k \geq 1$, $S_6(q)$ is vertex-transitive. ∎

Let G be a Chevalley group of normal type corresponding to the Lie algebra $\mathcal{L} = H \oplus \mathcal{L}^+ \oplus \mathcal{L}^-$, where H is the Cartan algebra and $\mathcal{L}^+$ ($\mathcal{L}^-$) is the direct sum of root subalgebras, corresponding to positive (negative) roots. The incidence graph $I(G)$ of the geometry $\gamma(G)$ of group G is a blow-up of the incidence graph $I(W)$ of the geometry of its Weyl group W (see [33,34]). In this case the blow-up $I(W)$ was constructed by using the Lie algebra $\mathcal{L}^+$ and a fixed Chevalley basis for it.

We restrict our attention to Chevalley groups of rank two of normal type. In this case we obtain a convenient description of graphs $S_n(q)$.

For each $b \in \mathcal{L}$, a linear transformation $ad(b) : x \mapsto [b, x]$ is a nilpotent operator of $\mathcal{L}$. Let $v = ad(te_\alpha)$, where e_α is an element of the Chevalley basis from the root space corresponding to root α, and $t \in \mathbb{F}_q$. Let $x_\alpha(t) = 1 + v/1! + v^2/2! + v^3/3! + \ldots$. Then $x_\alpha(t+t') = x_\alpha(t)x_\alpha(t')$, and G is generated by all $x_\alpha(t)$, $\alpha \in \mathcal{L}^+$, $t \in \mathbb{F}_q$. For a fixed positive root α, let U_α be a group generated by all $x_\alpha(t)$, $t \in \mathbb{F}_q$.

Proposition 2.2. *For $n = 3, 4, 6$, graph $S_n(q)$ is isomorphic to the incidence graph of the group incidence structure $\gamma = \gamma(U, \{U_{\alpha_1}, U_{\alpha_2}\})$.* ∎

Let $M_1 = \begin{pmatrix} 2 & -1 \\ -1 & 2 \end{pmatrix}$, $M_2 = \begin{pmatrix} 2 & -2 \\ -1 & 2 \end{pmatrix}$, $M_3 = \begin{pmatrix} 2 & -3 \\ -2 & 2 \end{pmatrix}$. (This is a complete list of the so-called 2×2 Cartan matrices.) In what follows A will represent a matrix from this list. We can consider a lattice $\mathcal{H}$ with basis $\{\alpha_1, \alpha_2\}$, i.e. the set $\{\lambda_1 \alpha_1 + \lambda_2 \alpha_2 \mid \lambda_1, \lambda_2 \in Z\}$. For an arbitrary 2×2 integer matrix $A = (a_{ij})$, we consider two linear transformations r_1, r_2 of $\mathcal{H}$, where $(\alpha_j)^{r_i} = \alpha_j - a_{ij}\alpha_i, i,j \in \{1,2\}$. It is easy to check that, if $A = M_k, k = 1, 2, 3$, then $r_i^2 = e, i = 1, 2$ and $(r_1 r_2)^m = e$ for $m = 3$ (if $k = 1$), $m = 4$ (if $k = 2$), $m = 6$ (if $k = 3$), and these conditions are generic relations for a group $W = W(A) = \langle r_1, r_2 \rangle$, i.e., $W(A)$ is isomorphic to the dihedral group D_m. $W(A)$ is usually called the *Weyl group* corresponding to the 2×2 Cartan matrix A. (For more on this, see [8].) The set $\Phi(A) = \{\alpha_i^g \mid g \in W, i = 1, 2\}$ is usually called a *root system*. The set $\Phi(A)$ is a disjoint union of sets $\Phi^+(A)$ and $\Phi^-(A)$, where $\Phi^+(A) = \Phi(A) \cap \{\lambda_1\alpha_1 + \lambda_2\alpha_2 \mid \lambda_i \geq 0, i = 1, 2\}$ (elements of $\Phi^+(A)$ are called *positive roots*) and $\Phi^-(A) = \Phi(A) \cap \{-x \mid x \in \Phi^+(A)\}$ (*negative roots*). We have $\Phi^+(M_1) = \{\alpha_1, \alpha_2, \alpha_1 + \alpha_2\}, \Phi^+(M_2) = \{\alpha_1, \alpha_2, \alpha_1 + \alpha_2, 2\alpha_1 + \alpha_2\}, \Phi^+(M_3) = \{\alpha_1, \alpha_2, \alpha_1 + \alpha_2, 2\alpha_1 + \alpha_2, 3\alpha_1 + \alpha_2, 3\alpha_1 + 2\alpha_2\}$. Let α_i^*, $i = 1, 2$, be the linear functional on $\mathcal{H}$ such that $\alpha_i^*(\alpha_j) = \delta_{ij}$, where δ_{ij} is the Kronecker delta. We can consider the dual lattice $\mathcal{H}^* = \{\lambda_1 \alpha_1^* + \lambda_2 \alpha_2^* \mid \lambda_1, \lambda_2 \in Z\}$. For a given $i, i = 1, 2, 3$, the group $W(A)$ acts on $\mathcal{H}^*$ by the following rule: for a linear functional $l \in \mathcal{H}^*$ and $g \in W$, $l \mapsto l^g$, where $l^g(x) = l(x^{g^{-1}})$ for all $x \in \mathcal{H}$.

Let $H_1(A) = \{(\alpha_1^*)^g \mid g \in W(A)\}$ and $H_2(A) = \{(\alpha_2^*)^g \mid g \in W(A)\}$. We shall say that two functionals $l_1 \in H_1(A)$ and $l_2 \in H_2(A)$ are incident and write $l_1 J l_2$ if and only if for every x from $\Phi(A)$, $l_1(x) \cdot l_2(x) \geq 0$.

Proposition 2.3. *The incidence structure $H(M_k) = (H_1(M_k), H_2(M_k), J)$ is isomorphic to the ordinary m_k-gon, $m_1 = 3, m_2 = 4, m_3 = 6$.*

PROOF.
It is easy to check that

$$H_1(M_1) = \{\alpha_1^*, -\alpha_1^* + \alpha_2^*, -\alpha_1^*\}$$
$$H_2(M_1) = \{\alpha_2^*, -\alpha_2^* + \alpha_1^*, -\alpha_2^*\}$$
$$H_1(M_2) = \{\alpha_1^*, -\alpha_1^* + 2\alpha_2^*, \alpha_1^* - 2\alpha_2^*, -\alpha_1^*\}$$
$$H_2(M_2) = \{\alpha_2^*, \alpha_1^* - \alpha_2^*, -\alpha_1^* + \alpha_2^*, -\alpha_2^*\}$$
$$H_1(M_3) = \{\alpha_1^*, \alpha_1^* - \alpha_2^*, 2\alpha_1^* - \alpha_2^*, -2\alpha_1^* + \alpha_2^*, -\alpha_1^* + \alpha_2^*, -\alpha_1^*\}$$
$$H_2(M_3) = \{\alpha_2^*, 3\alpha_1^* - \alpha_2^*, -3\alpha_1^* + 2\alpha_2^*, 3\alpha_1^* - 2\alpha_2^*, -3\alpha_1^* + \alpha_2^*, -\alpha_2^*\}$$

and the bipartite incidence graph for $H(M_k)$, $k = 1, 2, 3$, is the cycle C_{2m_k}. ∎

M. ore general propositions for $n \times n$ Cartan matrices of simple finite dimensional and affine algebras are considered in [34] and [35], respectively.

Let ζ be the vector space of formal linear combinations $t_1 \alpha_1^* + t_2 \alpha_2^*$, where $t_1, t_2 \in \mathbb{F}_q$, let $L = L(A)$ be the set of all linear combinations of the form $\sum_{\alpha \in \Phi^+(A)} t_\alpha e_\alpha, t_\alpha \in \mathbb{F}_q$, and let $\zeta \oplus L$ be the direct sum of ζ and L.

We define a bilinear product $[\,,\,]$ on $\zeta \oplus L$ by its values on elements of the basis in the following way:

(2.1)
$$\begin{cases} [\alpha_i^*, \alpha_2^*] = 0, \quad i = 1, 2 \\ [\alpha_i^*, e_\beta] = \alpha_i^*(\beta) \cdot e_\beta, \quad \beta \in \Phi^+(A), \quad i = 1, 2 \\ [e_\alpha, e_\beta] = \begin{cases} 0 & \text{if } \alpha + \beta \notin \Phi^+(A) \\ (r+1)e_{\alpha+\beta} & \text{if } \alpha + \beta \in \Phi^+(A), \end{cases} \end{cases}$$

where r is an integer uniquely determined by the condition $\beta - r\alpha \in \Phi(A)$, $\beta - (r+1)\alpha \notin \Phi(A)$.

It is known (see [11]) that $(\zeta \oplus L, [\,,\,])$ is isomorphic to the Borel subalgebra of the Lie algebra for G ($G = A_2(q), B_2(q), G_2(q)$), which is, by definition, the direct sum of the Cartan subalgebra with the sum of root spaces which correspond to positive roots.

Let us denote by L_α the totality of vectors in L of the form $\lambda\alpha, \lambda \in \mathbb{F}_q$. For an integer a we denote by $\bar{a}$ the residue of $a \pmod p$. We shall write $\bar{l} = \Sigma \bar{\lambda}_i \alpha_i^*$, where $l = \Sigma \lambda_i \alpha_i^*$ is an element of $\mathcal{H}^*$.

Let $l \in \mathcal{H}^*$ and $\eta(l) = \{\alpha \in \Phi^+ \mid l(\alpha) < 0\}$. We shall consider an incidence structure $\zeta(A, q)$ with the set of points and lines $\zeta_i(A, q) = \{(l, y) \mid l = l(x) \in H_i(A), y \in \sum_{\alpha \in \eta(l)} L_\alpha\}$, $i = 1, 2$, and the following incidence relation $\tilde{J}$:

$$(l(x), y) \tilde{J}(t(x), z) \iff lJt \text{ and } [\bar{l} + y, \bar{t} + z] = 0$$

It has been shown in [35,36] that the incidence structure of $\zeta(A, q)$ for $q = p^s$, $p > 3$, and $A \in \{M_1, M_2, M_3\}$ is isomorphic to the generalized n-gon arising from the Chevalley group $A_2(q), B_2(q), G_2(q)$, $n = 3, 4, 6$, respectively. If $A = M_2$, this statement is also valid for $p = 3$.

A mapping $\phi: (P, L, I) \to (P', L', I')$ is called a *morphism* from an incidence system (P, L, I) to an incidence system (P', L', I') if $\phi(P) \subset P'$, $\phi(L) \subset L'$, and pIl implies $\phi(p)I'\phi(l)$.

The following proposition follows immediately from definitions and the above.

Proposition 2.4. *A mapping $r : \zeta(M_k, q) \to H(M_k)$, $k = 1, 2, 3$, defined by $r((h, x)) = h$, is a morphism of the generalized m_k-gon onto the ordinary m_k-gon, $m_k \in \{3, 4, 6\}$.* ∎

If the characteristic of $\mathbb{F}_q$ is greater than 3, we can identify our graph $S_t(q)$, $t = 4, 6$, with the restrictions of the incidence graph for $\zeta(M_k, q), k = 2, 3$, on $r^{-1}(-\alpha_1^*) \cup r^{-1}(-\alpha_2^*)$. It is easy to see that $r^{-1}(-\alpha_i^*) = \{(-\alpha_i^*, x) \mid x \in \sum_{\alpha \in \eta(-\alpha_i^*)} L_\alpha\}$, $i = 1, 2$. Therefore $S_t(q)$ is the blow-up of $K_{1,1}$ with vertices $-\alpha_1^*$ and $-\alpha_2^*$.

Let $n = 4, 6$, $P_n = \{(x_1, x_2, \ldots, x_{n-1}) \mid x_i \in \mathbb{F}_q\}$, $L_n = \{[y_1, y_2, \ldots, y_{n-1}] \mid y_i \in \mathbb{F}_q\}$. We define an incidence relation I_n (between P_n and L_n) as:

$(a,b,c)I_4[x,y,z]$ if and only if

$$\begin{cases} y - b = xa \\ z - 2c = -2xb, \end{cases}$$

and $(a,b,c,d,f)I_6[x,y,z,u,w]$ if and only if

$$\begin{cases} y - b = xa \\ z - 2c = -2xb \\ u - 3d = -3xc \\ 2w - 3f = 3zb - 3yc - ua. \end{cases}$$

Theorem 2.5. *Let $q = p^m$, p is an odd positive prime. For $p \geq 3$, graph $S_4(q)$, and for $p \geq 5$, graph $S_6(q)$, are isomorphic to the incidence graphs of the incidence structure (P_4, L_4, I_4) and (P_6, L_6, I_6) respectively.* ∎

In order to prove Theorem 2.5, it is sufficient to represent elements of P and L by means of their coordinate vectors, and to choose coefficients in (2.1) in a certain way. Our choice is the following:

$$[e_{\alpha_1}, e_{\alpha_2}] = -e_{\alpha_1+\alpha_2}, \quad [e_{\alpha_1}, e_{\alpha_1+\alpha_2}] = -2e_{2\alpha_1+\alpha_2}$$
$$[e_{\alpha_1}, e_{2\alpha_1+\alpha_2}] = 3e_{3\alpha_1+\alpha_2}, \quad [e_{\alpha_2}, e_{3\alpha_1+\alpha_2}] = e_{3\alpha_1+2\alpha_2}$$
$$[e_{\alpha_1+\alpha_2}, e_{2\alpha_1+\alpha_2}] = 3e_{3\alpha_1+2\alpha_2}.$$

Let ϕ be the mapping of $S_6(q)$ to $S_4(q)$ induced by the canonical projections of vector spaces P_6 and L_6 on the first three coordinates. As an immediate corollary of Theorem 2.5, we get

Proposition 2.6. *ϕ is a morphism of the incidence systems.* ∎

1. . $S_3(q)$ is the incidence graph of the following incidence system: $P_3 = \{(x_1, x_2) \,|\, x_i \in \mathbb{F}_q\}$, $L_3 = \{[y_1, y_2] \,|\, y_i \in \mathbb{F}_q\}$, and $(x_1, x_2)I_3[y_1, y_2]$ if and only if $y_2 - x_2 = y_1 x_1$ (affine plane).

2. . Under the assumptions of Theorem 2.5, the operator $x_\alpha(t)$ from U preserves the set of vertices of $S_n(q)$, $n = 3, 4, 6$, and its restriction on this set is an automorphism of $S_n(q)$.

Let $H_n(q)$ be a blow–up of $K_{1,1}$ in the case when

(a) $(\mathcal{L}, *)$ is an n–dimensional algebra over $\mathbb{F}_q$ with a basis $\{e_1, e_2, \ldots, e_n\}$ and a multiplciation $*$ satisfying

(2.2) $$\begin{cases} e_i * e_1 = e_{i+1}, \quad e_1 * e_{i+1} = -e_{i+1}, \; i = 2, \ldots, n-1 \\ (1 \notin \{i,j\}) \Rightarrow e_i * e_j = 0 \end{cases}$$

(b) $\eta(p) = \{2, 3, \ldots, n\}$ and $\eta(l) = \{1, 3, \ldots, n\}$ for every point (p) and every line $[l]$.

Let $P_h(n) = P_h$ and $L_h(n) = L_h$ be the sets of points and lines of $H_n(q)$. It is easy to see, that point $(p) = (p_2, p_3, \ldots, p_n)$ and line $[l] = [l_1, l_3, \ldots, l_n]$ are incident if and only if the following conditions are satisfied

(2.3)
$$\begin{cases} l_3 - p_3 = l_1 \cdot p_2 \\ l_4 - p_4 = l_1 \cdot p_3 \\ \ldots\ldots\ldots\ldots \\ l_n - p_n = l_1 \cdot p_{n-1} \end{cases}$$

Theorem 2.7. *Graphs $H_n(q)$ are edge-transitive, and for $n \geq 3$, $g(H_n(q)) = 8$. Graph $S_4(q)$ is isomorphic to $H_3(q)$.*

Wenger [40] proved that $H_5(p)$ contains no C_{10}. His proof can be easily modified to obtain that $H_n(q)$, $q = p^m$, contains no C_{10}. This result, together with Theorem 2.7, implies

Proposition 2.8. *Graph $H_5(q)$ of order $2q^5$ and size q^6 contains no C_{10} and is not isomorphic to $S_6(q)$.* ∎

In fact, graph $H_5(q)$ does not contain C_{10} and, having girth 8, cannot be embedded into a generalized 6–gon. Other examples of "magnitude extremal" bipartite graphs of girth at most $2k - 2$, but containing no C_{2k}, can be found in [20].

3. Construction of graphs of order $2q^5$, size q^6, and girth 10 which is not based on a classical root system.

As we have mentioned, it was shown in [15] that there are no generalized 5–gons whose vertices have degree ≥ 3. This makes the construction of this section different from the one in Section 2.

Let P and L be two 5–dimensional vector spaces over the finite field $\mathbb{F}_q$. We assume that a basis in each of these spaces is chosen. Then the vectors of P and L can be thought as ordered 5–tuples of elements from $\mathbb{F}_q$. We define an incidence structure with point set P and line set L. It will be convenient for us to denote vectors from P as $x = (x) = (x_1, x_2, x_3, x_4, x_5)$ and vectors from L as $y = [y] = [y_1, y_2, y_3, y_4, y_5]$. The parentheses and brackets will allow us to distinguish vectors of different types (points and lines). We say that point $p = (p_1, \ldots, p_5)$ is incident with line $l = [l_1, \ldots, l_5]$, and we write it pIl or $(p)I[l]$, if and only if the following conditions are satisfied:

(3.1)
$$\begin{cases} l_2 - p_2 = l_1 p_1 \\ l_3 - p_3 = l_1 p_2 \\ l_4 - p_4 = p_1 l_2 \\ l_5 - p_5 = 2 l_1 p_4 - p_1 l_3 \end{cases}$$

This incidence defines a bipartite graph $B = B(q)$ whose vertex partition sets are P and L, and a point (p) and a line $[l]$ are connected by an edge if and only if $(p)I[l]$. The following theorem is the main result of this section.

Theorem 3.1. *The bipartite graph $B(q)$ satisfies the following properties:*
(a) $B(q)$ is q–regular of order $2q^5$ and size q^6
(b) For infinitely many values of q, $g(B(q)) = 10$
(c) For infinitely many values of q, $B(q)$ is not isomorphic to a subgraph of a generalized 6-gon.

PROOF.
(a) Obviously, $|V(B)| = |P| + |L| = q^5 + q^5 = 2q^5$. It is immediate from (3.1) that for a fixed $(p) \in V(B)$, the components of a line $[l] \in V(B)$ incident to (p) are determined uniquely by the value of l_1, which can be any element of the field. Therefore, the degree of (p) in B is q. In the same way we obtain that the degree of a line $[l]$ in B is also q. Therefore B is q–regular and $|E(B)| = q^6$.

Our proof of part (b) will be facilitated by the following two observations. First we notice that a graph G contains no $C_{2k}, k \geq 2$, if there is at most one simple path of length k between any two of its vertices. We will show that any pair of vertices of B is connected by at most one simple path of length $k, k = 2, 3, 4$. This will imply that $g(B) \geq 10$ since, being a bipartite graph, B contains no odd cycles.

Another observation is the existence of certain automorphisms of B. Let $x \in \mathbb{F}_q$, and $t_i(x), i = 0, \ldots, 5$, be the mappings $V(B) \to V(B)$ defined as

$$(p)^{t_0(x)} = (p_1 + x, p_2, p_3, p_4 + p_2 x, p_5 + 2p_3 x)$$
$$[l]^{t_0(x)} = [l_1, l_2 + l_1 x, l_3, l_4 + 2l_2 x + l_1 x^2, l_5 + l_3 x]$$
$$(p)^{t_1(x)} = (p_1, p_2 - p_1 x, p_3 - 2p_2 x + p_1 x^2, p_4, p_5 - p_4 x)$$
$$[l]^{t_1(x)} = [l_1 + x, l_2, l_3 - l_2 x, l_4, l_5 + l_4 x]$$
$$(p)^{t_2(x)} = (p_1, p_2 + x, p_3, p_4 - p_1 x, p_5 - 3p_2 x)$$
$$[l]^{t_2(x)} = [l_1, l_2 + x, l_3 + l_1 x, l_4, l_5 - 3l_2 x]$$
$$(p)^{t_3(x)} = (p_1, p_2, p_3 + x, p_4, p_5 + p_1 x)$$
$$[l]^{t_3(x)} = [l_1, l_2, l_3 + x, l_4, l_5]$$
$$(p)^{t_4(x)} = (p_1, p_2, p_3, p_4 + x, p_5)$$
$$[l]^{t_4(x)} = [l_1, l_2, l_3, l_4 + x, l_5 + 2l_1 x]$$
$$(p)^{t_5(x)} = (p_1, p_2, p_3, p_4, p_5 + x)$$
$$[l]^{t_5(x)} = [l_1, l_2, l_3, l_4, l_5 + x]$$

Lemma 3.2.
(i) For every $x \in \mathbb{F}_q$ and every $i \in \{0, 1, \ldots, 5\}$, the mapping $t_i(x)$ is an automorphism of the graph B and $t_i^{-1}(x) = t_i(-x)$.
(ii) For every edge $\{[l], (p)\}$ of B there exist automorphisms α and β of B such

that $[l]^\alpha = [0,0,0,0,0]$, $(p)^\alpha = (a,0,0,0,0)$, and $[l]^\beta = [b,0,0,0,0], (p)^\beta = (0,0,0,0,0)$, for some $a, b \in \mathbb{F}_q$. The automorphism group $Aut(B)$ acts transitively on the set of points and on the set of lines, and B is edge–transitive. ∎

Now we show that any pair of vertices of B is connected by at most one simple path of length 4. We need not distinguish between the two cases where both vertices are lines or points as the proofs are absolutely similar. So we assume that the two vertices are lines (it is also sufficient to consider this case only, if we want to show the absence of C_8 in B). Call the vertices $[l^1]$ and $[l^3]$. Let $[l^1]I(p^1)I[l^2]I(p^2)I[l^3]$ be our path. Due to Lemma 3.2 (ii), without loss of generality, we may assume $[l^1] = [0,0,0,0,0]$ and $(p^1) = (x,0,0,0,0)$. We denote the first components of $[l^2]$ and (p^2) by y and z correspondingly, and we write $[l^3]$ as $[a_1, a_2, a_3, a_4, a_5]$ (to avoid double indices). The conditions of adjacency of subsequent vertices of the path written in terms of their components (formula (3.1)) allow us to express all the components in terms of x, y, z, a_i: $(p^1)I[l^2]$ gives $[l^2] = [y, xy, 0, x^2y, 0]$ and $[l^2]I(p^2)$ gives $(p^2) = (z, y(x-z), -y^2(x-z), xy(x-z), -2xy^2(x-z))$. The last adjacency $(p^2)I[l^3]$, written in terms of components, gives

(3.2) $$\begin{cases} a_2 - y(x-z) = a_1 z \\ a_3 + y^2(x-z) = a_1 y(x-z) \\ a_4 - xy(x-z) = a_2 z \\ a_5 + 2xy^2(x-z) = 2a_1 xy(x-z) - a_3 z \end{cases}$$

We view (3.2) as a system of equations with unknown x, y, z and parameters a_i. The condition of existence of at most one simple path of length 4 between $[l^1]$ and $[l^3]$ is equivalent to the requirement that (3.2) has at most one solution which satisfies the following inequalities:

(3.3) $$\begin{cases} [l^1] \neq [l^3], [l^2] \neq [l^3], [l^1] \neq [l^2] \\ (p^1) \neq (p^2) \end{cases}$$

Simplifying, we get an equivalent system

(3.5) $$\begin{cases} [l^1] \neq [l^3], [l^2] \neq [l^3] \\ y \neq 0, x \neq z \end{cases}$$

Thus our goal is to prove that the combined system (3.2) and (3.4) has at most one solution for every $[l^3] = [a_1, a_2, a_3, a_4, a_5]$. The proof is not hard, it is completely elementary and we omit it.

Why does B contain no C_4 and no C_6? Due to Lemma 3.2, the existence of a cycle of length 4 in B would imply the existence of two interior vertex disjoint simple paths of length 2 between a pair of distinct lines $[l^1] = [0,0,0,0,0]$ and $[l^2] = [a_1, a_2, a_3, a_4, a_5]$. Let $(p) = (p_1, p_2, p_3, p_4, p_5)$ and $[l^1]I(p)I[l^2]$. Rewriting these adjacencies in terms of components, using (3.1), we get $p_2 = p_3 = p_4 = p_5 = 0$ and $a_2 = a_1 p_1$. If $a_1 \neq 0$, then p_1 is determined uniquely and, therefore, there exists only one path of length 2 between $[l^1]$ and $[l^2]$. If $a_1 = 0$, then $(p)I[l^2]$ implies $a_2 = a_3 = a_4 = a_5 = 0$, and $[l^1] = [l^2]$. Hence B contains no C_4.

Due to Lemma 3.2, the existence of a cycle of length 6 in B would imply the existence of two interior vertex disjoint simple paths of length 3 between a line $[l^1] = [0,0,0,0,0]$ and a point $(s) = (s_1, s_2, s_3, s_4, s_5)$. Let (p) and $[l^2]$ be two intermediate vertices on such a path, i.e., $[l^1]I(p)I[l^2]I(s), (p) \neq (s), [l^1] \neq [l^2]$. Rewriting the first two adjacencies in terms of components, we obtain $(p) = (x,0,0,0,0), [l^2] = [y, xy, 0, x^2y, 0]$ for some $x, y \in \mathbb{F}_q$. Using $[l^2]I(s)$, we obtain the following system:

(3.5) $$\begin{cases} xy - s_2 = s_1 y, 0 - s_3 = s_2 y, x^2 y - s_4 = s_1 xy \\ 0 - s_5 = 2s_4 y \\ (s) \neq (p), y \neq 0 \end{cases}$$

If $s_2 = 0$, then $s_3 = s_4 = s_5 = 0$ and $x = s_1$. This makes $(s) = (p)$, which is not the case. If $s_2 \neq 0$, then $s_3 \neq 0, y = -s_3/s_2$ and $x = s_1 - s_2^2/s_3$. Therefore (3.5) has no more than one solution with respect to x and y. Hence B contains no C_6, and $g(B) \geq 10$. To prove that B contains C_{10}, it is enough to show that there are two simple interior vertex–disjoint paths of length 5 between line $[l] = [0,0,0,0,0]$ and point $(p) = (0,1,1,1,1)$. This can be reduced to determining when the quadratic equation $3t^2 + 2t - 4 = 0$ has two distinct solutions which satisfy certain restrictions. It can be shown that for all sufficiently large values of q, which are neither divisible by 2 nor 3, and such that 13 is a quadratic residue in $\mathbb{F}_q$, such two solutions exist; the proof is straightforward and we omit it. We believe that $g(B) = 10$ for most other values of q (point (p) has to be chosen differently), but we leave this investigation out of the paper. This finishes the proof of part (b) of Theorem 3.1. Part (c) follows immediately from (b) since a generalized 6–gon has girth 12. ∎

Now we will construct graphs of order $2q^9$, size q^{10}, and girth ≥ 14. The construction is quite similar to the one we performed above. The same can be said about the logic of the proofs, though in this case some are shorter and more elegant. All proofs can be found in [18].

Let $P = \{(p) = (p_1, \ldots p_9) | p_i \in \mathbb{F}_q, i = 1, \ldots 9\}$ be the set of points and $L = \{[l] = [l_1, \ldots l_9] | l_i \in \mathbb{F}_q, i = 1, \ldots, 9\}$ be the set of lines. A point $(p) = (p_1, \ldots, p_9)$ and a line $[l] = [l_1, \ldots, l_9]$ are said to be incident (and denoted

$(p)I[l]$) if the following conditions are satisfied:

(3.6)
$$\begin{cases} l_2 - p_2 = l_1 p_1 \\ l_3 - p_3 = p_1 l_2 \\ l_4 - p_4 = l_1 p_2 \\ l_5 - p_5 = l_1 p_3 \\ l_6 - p_6 = p_1 l_4 \\ l_7 - p_7 = p_1 l_5 \\ l_8 - p_8 = l_1 p_6 \\ l_9 - p_9 = l_1 p_7 \end{cases}$$

This incidence defines a bipartite graph $G = G(q)$ whose vertex partition sets are P and L. It is easy to show that G has $2q^9$ vertices, q^{10} edges, and is q–regular.

For every $x \in \mathbb{F}_q$, we introduce the following mappings $t_i : V(G) \to V(G), i = 0, \ldots, 9$:

$(p)^{t_0(x)} = (p_1 + x, p_2, p_3 + p_2 x, p_4, p_5 + p_4 x, p_6, p_7 + p_6 x, p_8, p_9 + p_8 x)$
$[l]^{t_0(x)} = [l_1, l_2 + l_1 x, l_3 + 2l_2 x + l_1 x^2, l_4, l_5 + l_4 x, l_6 + l_4 x, l_7 + (l_5 + l_6)x + l_4 x^2,$
$\qquad l_8, l_9 + l_8 x]$
$(p)^{t_1(x)} = (p_1, p_2 - p_1 x, p_3, p_4 - 2p_2 x + p_1 x^2, p_5 - p_3 x, p_6 - p_3 x, p_7,$
$\qquad p_8 - (p_5 + p_6)x + p_3 x^2, p_9 - p_7 x)$
$[l]^{t_1(x)} = [l_1 + x, l_2, l_3, l_4 - l_2 x, l_5, l_6 - l_3 x, l_7, l_8 - l_5 x, l_9]$
$(p)^{t_2(x)} = (p_1, p_2 + x, p_3 - p_1 x, p_4, p_5 - p_2 x, p_6 + p_2 x, p_7 - p_3 x, p_8 + p_4 x, p_9 - p_5 x)$
$[l]^{t_2(x)} = [l_1, l_2 + x, l_3, l_4 + l_1 x, l_5 - l_2 x, l_6 + l_2 x, l_7 - l_3 x, l_8 + l_4 x, l_9 - l_5 x]$
$(p)^{t_3(x)} = (p_1, p_2, p_3 + x, p_4, p_5, p_6, p_7 + p_2 x, p_8, p_9 + p_4 x)$
$[l]^{t_3(x)} = [l_1, l_2, l_3 + x, l_4, l_5 + l_1 x, l_6, l_7 + l_2 x, l_8, l_9 + l_4 x]$
$(p)^{t_4(x)} = (p_1, p_2, p_3, p_4 + x, p_5, p_6 - p_1 x, p_7, p_8 - p_2 x, p_9)$
$[l]^{t_4(x)} = [l_1, l_2, l_3, l_4 + x, l_5, l_6, l_7, l_8 - l_2 x, l_9]$
$(p)^{t_5(x)} = (p_1, p_2, p_3, p_4, p_5 + x, p_6, p_7 - p_1 x, p_8, p_9 - p_2 x)$
$[l]^{t_5(x)} = [l_1, l_2, l_3, l_4, l_5 + x, l_6, l_7, l_8, l_9 - l_2 x]$
$(p)^{t_6(x)} = (p_1, p_2, p_3, p_4, p_5, p_6 + x, p_7, p_8, p_9)$
$[l]^{t_6(x)} = [l_1, l_2, l_3, l_4, l_5, l_6 + x, l_7, l_8 + l_1 x, l_9]$
$(p)^{t_7(x)} = (p_1, p_2, p_3, p_4, p_5, p_6, p_7 + x, p_8, p_9)$
$[l]^{t_7(x)} = [l_1, l_2, l_3, l_4, l_5, l_6, l_7 + x, l_8, l_9 + l_1 x]$
$(p)^{t_8(x)} = (p_1, p_2, p_3, p_4, p_5, p_6, p_7, p_8 + x, p_9)$
$[l]^{t_8(x)} = [l_1, l_2, l_3, l_4, l_5, l_6, l_7, l_8 + x, l_9]$
$(p)^{t_9(x)} = (p_1, p_2, p_3, p_4, p_5, p_6, p_7, p_8, p_9 + x)$
$[l]^{t_9(x)} = [l_1, l_2, l_3, l_4, l_5, l_6, l_7, l_8, l_9 + x]$

Lemma 3.3.
(i) For every $x \in \mathbb{F}_q$ and every $i \in \{1,\ldots 9\}$, the mapping $t_i(x)$ is an automorphism of the graph G and $t_i^{-1}(x) = t_i(-x)$.
(ii) For every edge $\{[l],(p)\}$ of G there exist automorphisms α and β of G such that $[l]^\alpha = [0,\ldots,0], (p)^\alpha = (a,0,\ldots,0)$, and $[l]^\beta = [b,0,\ldots,0], (p)^\beta = (0,\ldots,0)$, for some $a,b \in \mathbb{F}_q$. The automorphism group $Aut(G)$ acts transitively on each of the sets P and L, and G is edge–transitive. ∎

The proof of the following theorem is similar to the one of Theorem 3.1.

Theorem 3.4. *Let q be a prime power, $q \geq 3$. Then graph $G(q)$ is a q-regular bipartite graph of order $2q^9$ and girth ≥ 14. The automorphism group $Aut(G(q))$ is transitive on each of the sets of points and lines, and $G(q)$ is edge-transitive.* ∎

4. A family of graphs with large girth

In this section we construct a new infinite series of regular bipartite graphs with edge–transitive automorphism group and large girth. More precisely, for any positive odd integer $k \geq 3$ and any prime power q, we build a q–regular bipartite graph $D(k,q)$ on $2q^k$ vertices with girth $g \geq k+5$. Our construction generalizes the one of the graph $G(q)$ from Section 3, which turns out to be isomorphic to $D(9,q)$. Below we give several reasons why we find these graphs interesting.

As we mentioned in the Introduction, it is known that

$$ex(v, \{C_3, C_4, \cdots, C_m\}) \geq c_m v^{1+\frac{1}{m-1}},$$

for some positive constant c_m, $m \geq 3$, and the proof is nonconstructive. Graphs $D(k,q)$ demonstrate that for an infinite sequence of values of v, $ex(v, \{C_3, C_4, \cdots, C_{2s+1}\}) \geq d_s v^{1+\frac{1}{2s-3}}$, $s \geq 3$, and this is an improvement of the nonconstructive bound for large v. For large values of s and an infinite sequence of values of v, a better bound $ex(v, \{C_3, C_4, \cdots, C_{2s+1}\}) \geq f_s v^{1+\frac{4}{3}\frac{1}{2s+4}}$ is provided by some Ramanujan graphs (see below), and it appears to be the best asymptotic lower bound known. Comparing the exponents of v, we obtain that our bound is better for $3 \leq s \leq 11$ and large v. For all odd prime powers q or $q = 2^m$, m is a positive even integer, and all odd values of k, $3 \leq k \leq 17$, $k \neq 7$, graphs $D(k,q)$ are of the greatest known size among the graphs of given order and girth $\geq k+5$. The same is correct if $q = 2^m$, m is odd, k is odd, $3 \leq k \leq 17$, $k \neq 7, 11$. Graphs $D(3,q)$ and $D(5,q)$ have asymptotically as many edges as the incidence 'point–line' graphs of a generalized quadrangle and a generalized hexagon respectively, and the greatest known edge density (the ratio $e/\binom{v}{2}$) among the graphs of the same order and girth. For prime q, a somewhat similar construction leading to graphs with the same order, edge density and girth as $D(3,q)$ and $D(5,q)$, was done by Wenger (see [40] and Section 2.) Graph $D(7,q)$ has girth ≥ 12 but asymptotically fewer edges, than the incidence graph of a generalized hexagon whose girth is 12. Graph $D(9,q)$ is isomorphic to $G(q)$ from Section 3. It has girth at least 14 and shows that $ex(v, \{C_3, C_4, \cdots, C_{13}\}) \geq d_{13} v^{1+\frac{1}{9}}$. For $q = 2^m$, where m is an odd positive integer, this lower bound may not be the best due to a recent result of Ustimenko and Woldar[37,41], where an example of a q-regular graph of order $v = 2q^t$ and girth at least 16 is given, with t being an unknown integer satisfying the inequality $8 \leq t \leq 9$. Their result implies that $ex(v, \{C_3, C_4, \cdots, C_{15}\}) \geq d_{15} v^{1+\frac{1}{t}}$ for an infinite sequenc e of values of v and an integer t, $8 \leq t \leq 9$. This lower bound is certainly better than the one of magnitude $v^{1+\frac{1}{11}}$ provided by the graph $D(11,q)$.

Let $\{G_i\}$, $i \geq 1$, be a family of graphs such that each G_i is a k_i-regular graph of order v_i and girth g_i. Following Biggs [3] we say that $\{G_i\}$ is a family of graphs with *large girth* if

$$g_i \geq \gamma \log_{k_i - 1}(v_i)$$

for some constant γ. It is well known (e.g. see [6]) that $\gamma = 2$ would be the best possible constant, but no family has been found to achieve this bound. For many years the only significant results were the theorems of Erdös and Sachs and its improvements by Sauer, Walther, and others (see [6] pp. 107 for more details and references), who using nonconstructive methods proved the existence of infinite families with $\gamma = 1$. The first explicit examples of families with large girth were given by Margulis [23] with $\gamma \approx 0.44$ for some infinite families with arbitrary large valency, and $\gamma \approx 0.83$ for an infinite family of graphs of valency 4. The constructions were Cayley graphs of $SL_2(Z_p)$ with respect to special sets of generators. Imrich [16] was able to improve the result for an arbitrary large valency, $\gamma \approx 0.48$, and to produce a family of cubic graphs (valency 3) with $\gamma \approx 0.96$. In [5] a family of geometrically defined cubic graphs, so called sextet graphs, was introduced by Biggs and Hoare. They conjectured that these graphs have large girth. Weiss [38] proved the conjecture by showing that for the sextet graphs (or their double cover) $\gamma \geq 4/3$. Then independently Margulis [24,25,26] and Lubotsky, Phillips and Sarnak [21,22] (see also [28]) came up with similar examples of graphs with $\gamma \geq 4/3$ and arbitrary large valency (they turned out to be so-called Ramanujan graphs). In [4], Biggs and Boshier showed that γ is exactly $4/3$ for graphs from [22]. The graphs are Cayley graphs of the group $PSL_2(Z_q)$ with respect to a set of $p+1$ generators (p, q are primes congruent to 1 mod 4).

The family of graphs $D(k,q)$ presented below gives an explicit example of graphs with an arbitrary large valency and $\gamma \geq 1$. Their definition and analysis are basically elementary. All proofs are omitted, and can be found in [19]. Here are the graphs.

Let q be a prime power. We define the infinite semiplane $\Gamma(q)$ as follows. Let P and L be two infinite–dimensional vector spaces over the finite field F_q. The vectors of P and L can be thought as infinite sequences of elements of F_q. P and L will be the set of points and the set of lines of the incidence structure $\Gamma(q)$. It will be convenient for us to write the components of points and lines as

$$(p) = (p_1, p_{1,1}, p_{1,2}, p_{2,1}, p_{2,2}, p'_{2,2}, p_{2,3}, p_{3,2}, p_{3,3}, p'_{3,3}, \ldots, p'_{i,i}, p_{i,i+1},$$
$$p_{i+1,i}, p_{i+1,i+1}, \ldots),$$
$$[l] = [l_1, l_{1,1}, l_{1,2}, l_{2,1}, l_{2,2}, l'_{2,2}, l_{2,3}, l_{3,2}, l_{3,3}, l'_{3,3}, \ldots, l'_{i,i}, l_{i,i+1}, l_{i+1,i}, l_{i+1,i+1}, \ldots].$$

We also assume $p_{-1,0} = l_{0,-1} = p_{1,0} = l_{0,1} = 0$, $p_{0,0} = l_{0,0} = -1$, $p'_{0,0} = l'_{0,0} = 1$, $p_{0,1} = p_1$, $l_{1,0} = l_1$, $l'_{1,1} = l_{1,1}$, $p'_{1,1} = p_{1,1}$. We say that a point (p) is incident with a line $[l]$, and write it as $(p)I[l]$ if and only if the following conditions are satisfied:

(4.1)
$$\begin{cases} l_{i,i} - p_{i,i} = l_1 p_{i-1,i} \\ l'_{i,i} - p'_{i,i} = p_1 l_{i,i-1} \\ l_{i,i+1} - p_{i,i+1} = p_1 l_{i,i} \\ l_{i+1,i} - p_{i+1,i} = l_1 p'_{i,i} \\ \text{for} \quad i = 1, 2, \ldots \end{cases}$$

Notice that for $i = 1$, the first two equations coinside and give $l_{1,1} - p_{1,1} = l_1 p_1$. Let $D(q)$ be the incidence graph of the incidence structure $\Gamma(q) = (P, I, L)$. For an integer $k \geq 2$, let $\Gamma(k, q) = (P(k), I(k), L(k))$ be the incidence system, where $P(k)$ and $L(k)$ are images of P and L under the projection of these spaces on the first k coordinates, and $I(k)$ is defined by the first k equations of (4.1). (Actually we have $k - 1$ distinct equations, since for $i = 1$ the first two equation of the system (4.1) coincide.) Finally, let $D(k, q)$ be the incidence graph for $\Gamma(k, q)$.

Proposition 4.1. *Let $k \geq 2$. The incidence system $\Gamma(k, q)$ is a semiplane and $D(k, q)$ is a q-regular bipartite graph on $2q^k$ vertices containing no 4-cycles.* ∎

Our goal now is to show that the girth $g(D(k, q)) \geq k + 5$. This task will be greatly facilitated if we use some automorphisms of $D(k, q)$.

For every $x \in F_q$, let $t_1(x), t_2(x), t_{1,1}(x), t_{m,m+1}(x)$ and $t_{m+1,m}(x), m \geq 1$, $t_{m,m}(x)$ and $t'_{m,m}(x)$, $m \geq 2$, be maps of $P \to P$ and $L \to L$ defined by means of Table 1. An entry of the table shows the effect of the action of the corresponding map (top of the column) on the corresponding component of a line or a point (left end of the row). If the action of a map on the corresponding component of a point or a line is not defined by Table 1, it will mean that the component is fixed by the map. For example, the map $t_2(x)$ changes every component $l_{i,i+1}, i \geq 1$, of a line $[l]$ according to the rule : $l_{i,i+1} \to l_{i,i+1} + (l_{i,i} + l'_{i,i})x + l_{i,i-1}x^2$, and leaves every component $p_{i+1,i}, i \geq 1$, of a point (p) fixed; the map $t_{1,1}(x)$ changes every component $p_{i,i}, i \geq 1$, of a point (p) according to the rule $p_{i,i} \to p_{i,i} - p_{i-1,i-1}x$; the map $t_{5,6}(x)$ does not change components of any line $[l]$ (or any point (p)) which precede component $l_{5,6}$ (or $p_{5,6}$.)

Proposition 4.2. *For every $x \in F_q$, the maps $t_1(x), t_2(x), t_{1,1}(x); t_{m,m+1}(x)$ and $t_{m+1,m}(x), m \geq 1; t_{m,m}(x)$ and $t'_{m,m}(x), m \geq 2$, are automorphisms of $D(q)$, and their restrictions on $P(k) \cup L(k)$ are automorphisms of $D(k,q)$.* ∎

Finally, we summarize the properties of graphs $D(k,q)$ in the following two theorems.

Theorem 4.3. *For all integers $k \geq 2$ and all prime powers q, graphs $D(q)$ and $D(k,q)$ are edge-transitive. For $q = 2^n, n \geq 1$, and any even integer $k \geq 2$, graphs $D(q)$ and $D(k,q)$ are vertex-transitive.* ∎

Theorem 4.4. *Let $k \geq 3$ be a positive odd integer, q be a positive prime power, and $g = g(D(k,q))$ be the girth of graph $D(k,q)$. Then $g \geq k + 5$.* ∎

$i \geq 0$	$t_1(x)$	$t_2(x)$	$t_{11}(x)$	$t_{m,m+1}(x)$ $m \geq 1$	$t_{m+1,m}(x)$ $m \geq 1$	$t_{m,m}(x)$ $m \geq 1$	$t'_{m,m}(x)$ $m \geq 1$
$l_{i,i}$		$+l_{i,i-1}x$	$-l_{i-1,i-1}x$		$+l_{r,r-1}x,$ $r=i-m\geq 1$	$-l_{r,r}x,$ $r=i-m\geq 0$	
$l_{i,i+1}$		$+(l_{i,i}+l'_{i,i})x+$ $+l_{i,i-1}x^2$	$-l_{i-1,i}x$		$+l'_{r,r}x,$ $r=i-m\geq 0$	$-l_{r,r+1}x,$ $r=i-m\geq 0$	
$l_{i+1,i}$	$+l_{i,i}x$		$+l_{i,i-1}x$		$-l_{r,r}x,$ $r=i-m\geq 0$		$+l_{r+1,r}x,$ $r=i-m\geq 0$
$l'_{i,i}$	$+l_{i-1,i}x$	$+l_{i,i-1}x$	$+l'_{i-1,i-1}x$		$-l_{r-1,r}x,$ $r=i-m\geq 1$		$+l'_{r,r}x,$ $r=i-m\geq 0$
$p_{i,i}$	$+p_{i-1,i}x$	$+p_{i,i-1}x$	$-p_{i-1,i-1}x$	$+p_{r,r-1}x,$ $r=i-m\geq 1$		$-p_{r,r}x,$ $r=i-m\geq 0$	
$p_{i,i+1}$		$+p'_{i,i}x$	$-p_{i+1,i}x$	$+p'_{r,r}x,$ $r=i-m\geq 0$		$-p_{r,r+1}x,$ $r=i-m\geq 0$	
$p_{i+1,i}$	$+(p_{i,i}+p'_{i,i})x+$ $+p_{i-1,i}x^2$		$+p_{i,i-1}x$		$-p_{r,r}x,$ $r=i-m\geq 0$		$+p_{r+1,r}x,$ $r=i-m\geq 0$
$p'_{i,i}$	$+p_{i-1,i}x$		$+p'_{i-1,i-1}x$	$-p_{r-1,r}x,$ $r=i-m\geq 1$			$+p'_{r,r}x,$ $r=i-m\geq 0$

TABLE 1 $l_{0,0} = p_{0,0} = -1, l_{0,1} = p_{1,0} = 0; l_{1,0} = l_1; p_{0,1} = p_1; l'_{11} = l_{11}; p'_{11} = p_{11}; p'_{0,0} = l'_{0,0} = 1; p_{-1,0} = l_{0,-1} = p_{0,-1} = l_{-1,0} = 0.$

ACKNOWLEDGEMENTS

The authors are very grateful to Professors W.M. Kantor, G.A. Margulis, M. Simonovits and A. J. Woldar for conversations on the topics of this article. The observation that graph $H_5(p)$ contains an 8-cycle belongs to A. J. Woldar. Mr. C.S. Zack noticed the edge-transitivity of graphs $B(q)$ from Section 3.

References

1. C.T. Benson, *Minimal regular graphs of girths eight and twelve*, Canad. J. Math. **18** (1966), 1091–1094.
2. F. Bien, *Constructions of telephone networks by group representations*, Notices Amer. Math. Soc. **36** (1989), 5–22.
3. N.L. Biggs, *Graphs with large girth*, Ars Combinatoria **25–C** (1988), 73–80.
4. _____ and A.G. Boshier, *Note on the Girth of Ramanujan Graphs*, Journal of Combinatorial Theory, Series **B 49** (1990), 190–194.
5. _____ and M.J. Hoare, *The sextet construction for cubic graphs*, Combinatorica **3** (1983), 153–165.
6. B. Bollobás, *Extremal Graph Theory*, Academic Press, London, 1978.
7. J.A. Bondy and M. Simonovits, *Cycles of even length in graphs*, J. Combinatorial Theory (B) **16** (1974), 97–105.
8. N. Bourbaki, *Groupes et algèbras de Lie*, Chap. IV, V, VI, Hermann, Paris, 1968.
9. A.E. Brouwer, A.M. Cohen, A. Neumaier, *Distance – Regular Graphs*, Springer–Verlag, Heidelberg–New York, 1989.
10. W.G. Brown, *On graphs that do not contain a Thomsen graph*, Canad. Math. Bull. **9** (1966), 281–285.
11. R.W. Carter, *Simple Groups of Lie Type*, Wiley, New York, 1972.
12. F. K. Chung, *Constructing random–like graphs*, Probabilistic Combinatorics and its Applications Lecture Notes, A.M.S., San Francisco, 1991, pp. 1–24.
13. P. Erdös, A. Rényi and V.T. Sós, *On a problem of graph theory*, Studia Sci. Math. Hungar. **1** (1966), 215–235.
14. R. J. Faudree and M. Simonovits, *On a class of degenerate extremal graph problems*, Combinatorica **3 (1)** (1983), 83–93.
15. W. Feit and G. Higman, *The non–existence of certain generalized polygons*, J. Algebra **1** (1964), 114–131.
16. W. Imrich, Explicit construction of graphs without small cycles, C ombinatorica **2** (1984), 53–59.
17. W.M. Kantor, Generalized polygons, SCABs and GABs In Buildings and the Geometry of Diagrams, Lecture Notes in Math **1181** (1986), Springer Verlag, Berlin, 79–158.
18. F. Lazebnik, V. A. Ustimenko, *New examples of graphs without small cycles and of large size*, submitted for publication.
19. _____, *Explicit construction of graphs with an arbitrary large girth and of large size*, submitted for publication.
20. _____ and A. J. Woldar, *Properties of Certain Families of 2k–Cycle Free Graphs*, submitted for publication.
21. A. Lubotsky, R. Phillips, P. Sarnak, *Ramanujan conjecture and explicit constructions of expanders*, Proc. Stoc **86** (1986), 240–246.
22. _____, *Ramanujan graphs*, Combinatorica **8 (3)** (1988), 261–277.
23. G.A. Margulis, *Explicit construction of graphs without short cycles and low density codes*, Combinatorica **2** (1982), 71–78.
24. _____, *Arithmetic groups and graphs without short cycles*, 6th Internat. Symp. on Information Theory, Tashkent, Abstracts, Vol . 1, 1984, pp. 123–125. (Russian)
25. _____, *Some new constructions of low–density paritycheck codes*, 3rd Internat. Seminar on Information Theory, convolution codes and multy–user communication, Sochi (1987), . 275–279. (Russian)
26. _____, *Explicit group–theoretical construction of combinatori al schemes and their application to the design of expanders and concentrators*, Problemy Peredachi Informatsii **24, No. 1**, 51–60; English transl. publ Journal of Problems of Information Transmission (1988), 39–46.
27. S. Payne and J.A. Thas, *Finite generalized quadrangles*, Pitman, New York, 1985.
28. P. Sarnak, *Some applications of modular forms*, Cambridge Tracts in Mathematics **99** (1990), Cambridge Univ. Press.

29. M. Simonovits, *Extermal Graph Theory*, Selected Topics in Graph Theory 2 (L.W. Beineke and R.J. Wilson, eds.), Academic Press, London, 1983, pp. 161–200.
30. R.R. Singleton, *On minimal graphs of maximum even girth*, J. Combinatorial Theory **1** (1966), 306–322.
31. J. Tits, *Sur la trialité et certains groupes qui s'en déduisent*, Publ. Math. I.H.E.S. **2** (1959), 14–20.
32. _____, *Buildings of spherical type and finite BN-pairs*, Lecture Notes in Math **386**, Springer–Verlag, Berlin, 1974.
33. V.A. Ustimenko, *Division algebras and Tits geometries*, DAN USSR **296, No. 5** (1987), 1061–1065. (Russian)
34. _____, *A linear interpretation of the flag geometries of Chevalley groups*, Kiev University, Ukrainskii Matematicheskii Zhurnal **42, No. 3** (March, 1990), 383–387; English transl..
35. _____ paper On the embeddings of some geometries and flag systems in Lie algebras and superalgebras, Root systems, representation and geometries (1990), Kiev, IM AN UkrSSR, 3–16.
36. _____, *On some properties of geometries of the Chevalley groups and their generalizations*, Studies in Algebraic Theory of Combinorial Objects (1986), Moskow; English transl (1991), Kluwer Publ., Dordresht, 112–121.
37. _____ and A.J. Woldar, *An improvement on the Erdös bound for graphs of girth 16*, submitted for publication.
38. A.I. Weiss, *Girth of bipartite sextet graphs*, Combinatorica **4 (2–3)** (1984), 241–245.
39. R. Weiss, *Distance-transitive graphs and generalized polygons*, Arch. Math. **45** (1985), 186–192.
40. R.Wenger, *Extremal graphs with no C^4, C^6, or C^{10}'s*, J. of Combinatorial Theory **Series B 52** (1991), 113-116.
41. A. J. Woldar and V. A. Ustimenko, *An application of group theory to extremal graph theory*, submitted for publication.

DEPARTMENT OF MATHEMATICAL SCIENCES, UNIVERSITY OF DELAWARE, NEWARK, DELAWARE 19716

E-mail address: fellaz@gluttony.math.udel.edu

DEPARTMENT OF MATHEMATICS AND MECHANICS, KIEV STATE UNIVERSITY, 6 GLUSHKOV PROSPECT, KIEV 252127, THE UKRAINE

E-mail address: vau%drp.univ.kiev.ac@r4elay.ussr.eu.net

GROUPS AND EXPANDERS

A. LUBOTZKY AND B. WEISS

October 1992

ABSTRACT. Most explicit examples of expanding graphs are families of Cayley graphs of finite groups. In this note we consider the question of what is it that makes a family of such Cayley graphs expanders- is it a property of the groups alone or does it also depend upon the set of generators? We also give some conditions which prevent an infinite family of groups from being expanders, discuss some examples and raise some further problems and conjectures.

1. INTRODUCTION

Expanding graphs have received a great deal of attention in recent years, mainly because of their importance in computer science and in particular their usefulness in constructing communication networks. This interest led to various explicit constructions of families of expanders ([M1], [LPS], [M2], [L1], [C], [Mo]). These constructions usually take the following from. Starting with an infinite group Γ and finite set of generators Σ, let $\pi_i : \Gamma \to G_i$ be an infinite family of finite quotients. The Cayley graphs $\{X(G_i, \pi_i(\Sigma))\}$ will be the family of expanders if the right conditions prevail. The basic reasons vary, one uses property T of Γ, Selberg's theorem $\lambda_1 \geq \frac{3}{16}$, the Ramanujan conjecture or one of their modifications. We give a brief survey in §2.

In §3, we give some general methods of constructing families of bounded degree Cayley graphs that are not expanders. For example: theorem 3.1 states that if Γ is a finitely generated amenable group, $\Gamma = \langle \Sigma \rangle$, then the Cayley graphs $\{X_i = X(\pi_i(\Gamma), \pi_i(\Sigma))\}$ are non expanders for any infinite family of finite quotients. That means that for every $\varepsilon > o$, there is some i such that X_i is not an ε-expander.

As a corollary we deduce in 3.2 that if $\{G_i\}$ is an infinite family of finite solvable groups of bounded derived length, and $G_i = \langle \Sigma_i \rangle$ with the cardinality of the Σ_i bounded, then $\{X(G_i, \Sigma_i)\}$ is a non expander family. In contrast to this we give an example in 3.3 of an infinite family of solvable groups which are expanders thus answering a question raised in [AB].

1991 *Mathematics Subject Classification.* Primary 05C25 ; Secondary 22E40.

Research of the second author was partially sponsored by the Edmund Landau Center for research in Mathematical Analysis, supported by the Minerva Foundation (Germany).

This paper is in final form and no version of it will be submitted for publication elsewhere.

The results of §2 and §3 suggest a very basic question: does the expander - non expander property of a family of groups depend on the groups alone or is it also a function of the generating sets. More precisely we formulate

Problem 1.1. Let $\{G_i\}$ be a family of finite groups, $\langle \Sigma_i \rangle = \langle \Sigma_i' \rangle = G_i$ and $|\Sigma_i|$, $|\Sigma_i'| \leq k$, for all i. Does the fact that $\{X(G_i, \Sigma_i)\}$ is an expander family imply the same for $\{X(G_i, \Sigma_i')\}$?

The known methods of proving that a family is an expander certainly depend on the choice of generators. For example the groups $SL_n(p)$ for fixed n and p a prime are known to be expanders with the generators chosen globally (from $SL_n(\mathbb{Z})$). Computations by Laffery and Rockmore [LR] suggest (for $n = 2$) that randomly chosen generators, or possible even "worst case" generators for these groups also give expanders.

A positive answer to problem 1.1 would in fact be surprising. Some evidence for such an answer is the following. The natural strategy to 1.1 is to find a profinite group K with two finitely generated dense subgroups, A and B, with A amenable and B having property T. Then the finite quotients $\{G_i\}$ of K would give a negative answer to 1.1. We conjecture that this is impossible.

Conjecture 1.2. Let K be a compact group. If A and B are both finitely generated subgroups that are dense in K with A amenable and B having property T then K is finite.

The latter part of the paper is devoted mainly to some evidence favoring this conjecture. We show in 5.5. that this conjecture is valid if K is a linear group over some field.

We also observe that while for $n \geq 3$

$$K_n = \prod_p SL_n(p)$$

has a dense subgroup with property T it doesn't have a dense finitely generated amenable group. In contrast with this we consider in §4

$$K^p = \prod_n SL_n(p)$$

and show that it does contain a finitely generated dense amenable group. Some standard conjectures on the congruence subgroup problem imply that all known examples of groups with property T **cannot** be densely embedded in K^p. Thus both K_n and K^p seem to support our conjecture for very different reasons.

2. Expander groups

A finite k-regular graph X is called an ε-**expander** ($\varepsilon > 0$) if for all subsets $A \subset X$ with $|A| \leq \frac{1}{2}|X|$ we have $|\partial A| \geq \varepsilon |A|$ where

$$\partial A = \{y \in X : \text{distance}(y, A) = 1\}.$$

We will say that a family of groups $Y = \{G_i\}$ is an **expander** family if for some k and $\varepsilon > 0$ there are generating sets Σ_i for G_i of size at most k such that all the Cayley graphs $X(G_i, \Sigma_i)$ are ε-expanders. The family Y will be called a **non-expander** family if for some k there are generating sets Σ_i for G_i of size at most k, and for all positive ε at least one of the Cayley graphs $X_i = X(G_i, \Sigma_i)$ is not an ε-expander.

The major problem that motivated the work presented here is whether or not a family of groups can be both an expander family and a non-expander family. In this section we will survey the known methods of constructing, of finding, expander families, while the next section will be devoted to non-expander families.

The known expander families of groups have the following common structure. Start with an infinite group Γ that is finitely generated and fix a set of generators Σ. The family $\{G_i\}$ is a family of homomorphic images of Γ, $\pi_i(\Gamma)$ with $\Sigma_i = \pi_i(\Sigma)$. Here are the main expander families. (The reader is referred to [L1] for more details.)

I: Let Γ be a countable group with property T (see [M3] or [L1] for definitions) Σ any finite generating set, and take for G_i all finite quotient groups of Γ. This result is due to Margulis in [M1]. If Γ is a lattice in a semi simple Lie group all of whose factors H_j have rank ≥ 2 then Γ has property T (the H_j can be Lie groups over an arbitrary local field).

A weaker property, called τ, suffices for the above to hold, see [LZ]. A group Γ has property τ if all of the finite dimensional unitary representations of Γ are bounded away from the trivial representation. For this it suffices that Γ be an irreducible lattice in $\prod_j H_j$, with the H_j simple Lie groups and at least **one** of the H_j's having rank ≥ 2.

II: Let M be a compact Riemannian manifold, $\Gamma = \pi_1(M)$ and N_i a family of finite index normal subgroups of Γ such that the corresponding finite covers M_i of M satisfy $\lambda_1(M_i) \geq c > 0$ where $\lambda_1(M_i)$ is the smallest positive eigenvalue of the Laplacian on M_i. Assume Σ is a finite set of generators for Γ and take for π_i the canonical projections $\pi_i : \Gamma \to \Gamma/N_i$. Instead of compactness one can assume $M = X/\Gamma$ with X a symmetric space and Γ any lattice in $\text{Aut}(X)$. An example of this situation is when M is the modular surface, $\Gamma = SL_2(\mathbb{Z})$ and N_i the congruence subgroups of Γ, i.e. the G_i are $SL_2(\mathbb{Z}/i\mathbb{Z})$. The fact that one can take here $c = 3/16$ is a well known result of Selberg [S]. The Jacquet-Langlands correspondence extends this to uniform arithmetic lattices of $SL_2(\mathbb{R})$ as well.

Selberg's result was extended to all arithmetic groups in $SO(n,1)$ ([EGM], [LPSS]) and $SU(n,1)$ ([Li]). From these results it can be shown that one gets an expander family for every characteristic zero S-arithmetic group (with N_i the congruence subgroups) provided that Γ is an ineducable lattice in $\prod H_j$ and at least one of the H_j's is a real simple Lie group of rank≥ 1.

III: Let Γ be an arithmetic lattice in $SL_2(Q_p)$ and N_i the family of conguvence subgroups, Σ any finite set of generators. The fact that we get a family of expanders follows now from the Ramanujan conjecture proved by Eichler and Deligne. This is shown in [LPS] and [M2]. Here Q_p may be replaced by any non-Archimedean local field of arbitrary characteristic. For positive characteristic this follows from the work of Drinfeld who proved the analogue of the Ramanujan conjecture there. This was shown in [Mo], (as explained there the Jacauet-Langlands correspondence is needed once again).

Given all of the above the following conjecture suggests itself:

Conjecture 2.1. Let k be a global field, S a finite set of primes (i.e. valuations) of k that contains all Archimedean ones. Let G be a simple, simply connected, (connected) algebraic group defined over k and $\Gamma = G(O_S)$ where O_S is the ring of S-integers of k, i.e.

$$O_S = \{x \in k : \nu(x) \geq 0 \text{ for all } \nu \notin S\}.$$

Assume that Γ has a finite generating set Σ and that N_i are normal congruence subgroups. Then with $\pi_i : \Gamma \to \Gamma/N_i$ the canonical projections the Cayley graphs $X(\pi_i(\Gamma), \pi_i(\Sigma))$ are ε-expanders for some fixed $\varepsilon > 0$.

Remark 2.2.

- **(i):** In almost all cases Γ is indeed finitely generated, the only exception is when $\text{char}(k) = p > 0$ and $\sum_{\nu \in S} \text{rank}(G(k_\nu)) = 1$ (see [Be] or [L2])
- **(ii):** In most cases the conjecture is known to hold. Indeed Γ is an irreducible lattice in $\prod_{\nu \in S} G(k_\nu)$, where k_ν is the completion of k with respect to ν. If for some ν, $\text{rank}(G(k_\nu))$ is at least 2 then the conjecture holds as explained in (I). If for all ν, $\text{rank}(G(k_\nu)) \leq 1$ but one of these k_ν is $\mathbb{R}$ or $\mathbb{C}$ then the conjecture holds by (II) above. If at least one of the $G(k_\nu)$ is $SL_2(k_\nu)$ the conjecture holds by (III). In order to prove it in full generality it would suffice to work out the analogues of [EGM] (or [LPSS]) and [Li] for rank one groups (not SL_2) over non-Archimedean fields.

3. Non expander groups

In the preceding section we surveyed the methods of finding expander families of groups. Here we show how to find non expander families. In many ways the opposite of groups with property T are amenable groups. As we shall see this is also the case with respect to the expander property. Recall that a group Γ is amenable if $\ell^\infty(\Gamma)$ has a Γ-invariant mean. By a well known result of Følner (cf. [L1]) and the references there) a discrete group is amenable if and only if for every finite set Σ and every $\delta > 0$, there is a finite set $A \subset \Gamma$ satisfying

$$|\Sigma A \Delta A| < \varepsilon |A|$$

where $\Sigma A = \{sa : s \in \Sigma, a \in A\}$ and Δ denotes the symmetric difference. In words, A is "almost invariant" with respect to left multiplication by Σ.

Theorem 3.1. *Let Γ be a finitely generated amenable group and $Y = \{G_i\}$ an infinite family of quotient groups, then Y is a non expander family.*

Remark 3.2. If $\pi_i : \Gamma \to G_i$ are the canonical projections and Σ is **any** generating set for Γ we shall show that the Cayley graphs $X(G_i, \pi_i(\Sigma))$ are not an ε-expander family for any $\varepsilon > 0$. This is a little more than the assertion that Y is a non expander family.

Proof. Fix a generating set Σ, and a positive $\varepsilon > 0$. Find a finite set A in Γ such that $|sA \Delta A| < \varepsilon |A|$ for all $s \in \Sigma \cup \Sigma^{-1}$. If $\pi : \Gamma \to G_i$ is onto one of our quotient groups and $L_i = \ker(\pi)$, define a function on G_i by

$$\varphi(g) = \sum_{h \in \pi^{-1}(g)} 1_A(h).$$

Calculating

(1)
$$\begin{aligned}
\|\pi(s)\varphi - \varphi\|_{\ell_1} &= \sum_g |\varphi(\pi(s)g) - \varphi(g)| \\
&= \sum_g \left| \sum_{h \in \pi^{-1}(g)} 1_A(sh) - \sum_{h \in \pi^{-1}(g)} 1_A(h) \right| \\
&\leq \sum_{h \in \Gamma} |1_A(sh) - 1_A(h)| \\
&= |A \Delta sA| + |s^{-1}A \Delta A| < 2\varepsilon|A| \\
&= 2\varepsilon\varepsilon \|\varphi\|_{\ell_1}.
\end{aligned}$$

Thus φ is an approximately invariant ℓ_1 function on G_i. Furthermore, since there are infinitely many G_i's it is clear that we may assume that φ is far from being a constant function, indeed it's support is at most of size $|A|$ and we may assume $|G_i| > 2|A|$.

To conclude the proof we repeat the standard argument that an approximately invariant function gives rise to approximately invariant sets which will preclude the possibility that $\{G_i\}$ is an $\sqrt{2\varepsilon}$-expander family.

Define
$$B_j = \{g \in G_i : \varphi(g) \geq j\}.$$

Clearly, if 1_{B_j} are the indicator functions of the B_j's we have $\varphi = \sum_{j=1}^{\infty} 1_{B_j}$, and one can also easily check that

(2)
$$\|\pi(s)\varphi - \varphi\|_{\ell_1} = \sum_{j=1}^{\infty} \|\pi(s)1_{B_j} - 1_{B_j}\|_{\ell_1}$$

Denote by
$$J_s = \{j : \|\pi(s)1_{B_j} - 1_{B_j}\|_{\ell_1} > \sqrt{2\varepsilon}\|1_{B_j}\|_{\ell_1}\}$$

and compute

$$\sqrt{2\varepsilon} \sum_{j \in J_s} \|1_{B_j}\|_{\ell_1} < \sum_{j \in J_s} \|\pi(s)1_{B_j} - 1_{B_j}\|_{\ell_1} \leq \|\pi(s)\varphi - \varphi\|_{\ell_1}$$

$$\leq 2\varepsilon \|\varphi\|_{\ell_1}$$

by (1) and (2), hence

(3)
$$\sum_{j \in J_s} \|1_{B_j}\|_{\ell_1} \leq \sqrt{2\varepsilon} \|\varphi\|_{\ell_1}$$

Since $\|\varphi\|_{\ell_1} = \sum_1^{\infty} \|1_{B_j}\|_{\ell_1}$, if $|\Sigma| \cdot \sqrt{2\varepsilon} < 1$ there are indices j_0 that are not in any of the J_s's and any such B_{j_0} is clearly $\sqrt{2\varepsilon}$-invariant under all the $\pi(s)$, $s \in \Sigma$. Since $\varepsilon > 0$ was arbitrary we have shown that $\{G_i\}$ is a non-expander family. $\square$

For a group G, denote $G_1 = G$ and $G_{i+1} = [G_i, G_i]$. Recall that G is solvable of derived length ℓ if $G_\ell \neq \{e\}$ but $G_{\ell+1} = \{e\}$.

Corollary 3.3. *Let $Y = \{G_i\}$ be an infinite family of finite solvable groups that are all of derived length at most ℓ and are also generated by k elements, say Σ_i generates G_i, $|\Sigma_i| = k$ for all i. Then the Cayley graphs $X(G_i, \Sigma_i)$ are not a family of expanders.*

Proof. Let $F = F(x_1, \ldots, x_k)$ be the free group on k-generators, $\Sigma = \{x_1, \ldots, x_k\}$. The group $\Gamma = F/F_{\ell+1}$ is a solvable group and the map π_i taking Σ to Σ_i extends to a homomorphism from Γ onto G_i. Since Γ is amenable the corollary follows at once from the theorem. □

The bound on the derived length is essential as the following example of an infinite expander family of solvable groups shows:

Example 3.4. Let $\Gamma = SL_n(\mathbb{Z})$ and for a fixed prime p set
$$\Gamma(p^m) = \ker(SL_n(\mathbb{Z}) \to SL(\mathbb{Z}/p^m\mathbb{Z})).$$
As is well known, Γ is finitely generated, for example if $\{e_1, \ldots, e_n\}$ is the standard basis of $\mathbb{R}^n$, and T, S are defined by
$$Te_i = \begin{cases} e_1 + e_2 & i = 1 \\ e_i & i \neq 1 \end{cases}$$
$$Se_i = \begin{cases} e_{i+1} & i < n \\ (-1)^{n-1}e_1 & i = n \end{cases}$$
then T and S generate Γ. Since $\Gamma(p)$ is of finite index in Γ it too is finitely generated, say by a finite set Σ. Now the Cayley graphs $X_m = X(\Gamma(p)/\Gamma(p^m), \Sigma)$ satisfy:

(i): $\Gamma(p)/\Gamma(p^m)$ is a p-group and hence solvable (it is even nilpotent) for all m.
(ii): For some $\varepsilon > 0$, the X_m are all ε-expanders.

To check (i) one verifies easily that the order of $\Gamma(p)/\Gamma(p^m)$ is $p^{(m-1)(n^2-1)}$. For (ii) we distinguish between $n \geq 3$ and $n = 2$. In the first case $\Gamma(p)$ is a lattice in $SL_n(\mathbb{R})$ and thus has property (T) and by I. of §2 we have (ii). The result for the second case follows from Selberg's theorem as we have already indicated in II of §2.

This example answers a question raised in [AB].

We turn now to another method for showing that a family of groups cannot be made into a family of expanders with respect to any (bounded) set of generators. First a definition;

Definition 3.5. Let G be a group together with a finite generating set Σ. For a fixed unitary representation ρ denote by
$$K(G, \Sigma, \rho) = \inf_{||v||=1} \max_{s \in \Sigma} ||\rho(s)v - v||^2$$
and let
$$K(G, \Sigma) = \inf\{K(G, \Sigma, \rho) : \rho \text{ a representation}$$
$$\text{that contain no non zero fixed vectors}\}$$

This will be called the Kazhdan constant of (G, Σ), and a group has property T if and only if this constant is positive for some (and hence all) finite generating set.

Theorem 3.6. *For any $k > 0$ and positive d, there is a constant $c = c(k, d)$ such that if G is any group with generators Σ satisfying*

$$(a)\ |\Sigma| \leq d; \quad (b)\ K(G, \Sigma) \geq k$$

then for any subgroup $G_0 < G$ with $[G : G_0] \leq n$

$$|G_0/[G_0, G_0]| < c^n.$$

Before proving this rather technical theorem here is a corollary that fulfills the promise that we just made about more non expander families.

Corollary 3.7. *1. If $X_i = X(G_i, \Sigma_i)$ is a family of Cayley graphs with $|\Sigma_i|$ bounded that is an expander, then there is a constant $c > 0$ such that for all i, and any subgroup M of index n in G_i*

$$|M/[M, M]| \leq c^n.$$

2. Let M_i be any family of finite groups, $\{p_i\}$ an infinite sequence of primes and let

$$G_i = C_{p_i} wr M_i$$

denote the wreath product of the cyclic group with p_i elements and M_i, i.e.

$$G_i = F_{p_i}[M_i] \rtimes M_i$$

when $F_{p_i}[M_i]$ is the additive group of the group algebra of M_i over the field F_{p_i} and M_i acts on it by multiplication. Then $X(G_i, \Sigma_i)$ is not an expander family with respect to any bounded set of generators Σ_i.

Proof. 1. It is shown in [L1,Chap. 4] (cf. also [LZ]) that $X(G_i, \Sigma_i)$ are an expander family if and only if $K(G_i, \Sigma_i)$ is bounded away from zero. Thus this follows at once from theorem 3.5.

2. This follows from 1. since G_i contains an abelian subgroup $A_i = F_{p_i}[M_i]$ which is of index $|M_i|$ but of size $p_i^{|M_i|}$ and hence for all c, if p_i is sufficiently large $|A_i/[A_i, A_i]| < c^{[G_i:A_i]}$ cannot hold. □

We do not know that the corollary continues to hold if the p_i are a fixed prime, although we do expect it to. An interesting special case is $F_2[S_n] \rtimes S_n$ where S_n is the symmetric group on n letters.

Remark 3.8. If Γ has property T then it cannot have finite index group with infinite abelian quotients. Our theorem 3.5 may be viewed as a quantitative version of this giving for any such Γ an estimate

(4) $$|H/[H, H]| \leq c^{[\Gamma:H]}$$

for some constant c, since as is well known groups with property T are finitely generated. A polynomial bound in (4), suggested by the known examples of groups with property T, would have interesting applications.

For the proof of theorem 3.5 we need the following:

Proposition 3.9. *Let G be a group generated by Σ, $|\Sigma| = d$, and H a finite index subgroup. Then H has a generating set Σ' satisfying*

   (a): $|\Sigma'| \leq d \cdot [G : H]$
   (b): $K(H, \Sigma') \geq K(G, \Sigma)$.

Proof. Denote by n the index $[G : H]$ and let $T = \{t_1, \ldots, t_n\}$ be a right transversal for H in G, i.e. right coset representatives and denote by $g \to \bar{g}$ the map from $G \to T$ such that
$$g\bar{g}^{-1} \in H.$$
As is well known H is generated by Σ' the non trivial elements of the form $(t_i s)(\overline{t_i s})^{-1}$, $t_i \in T$, $s \in \Sigma$ (cf. [MKS, p.89]). For any unitary representation ρ of H on a Hilbert space V with no non zero fixed vector let $(\tilde{V}, \tilde{\rho}) = \text{Ind}_H^G(\rho)$ denoted the induced representation. Recall that $\tilde{V}$ is the space of functions $f : G \to V$ that satisfy
$$f(hg) = \rho(h)f(g)$$
with G acting on the right. Such a function is determined as soon as $f(t_i)$, $t_i \in T$ is given and hence $\tilde{V}$ may be identified with V^n as the direct sum of n-copies of V. One checks directly or via Frobenius reciprocity that $\tilde{\rho}$ doesn't contain the trivial representation.

Now given $v_0 \in V$, define $f \in \tilde{V}$ by setting $f(t_i) = v_0/\sqrt{n}$ for $i = 1, \ldots, n$. Then $\|f\| = 1$ and therefore by the definition of $K(G, \Sigma)$ there is some $s \in \Sigma$ such that

$$\begin{aligned}
K(G, \Sigma) \leq \|\tilde{\rho}(s)f - f\|^2 &= \sum_{i=1}^{n} \|f(t_i s) - f(t_i)\|^2 \\
&= \sum_{i=1}^{n} \|f(t_i s (\overline{t_i s})^{-1} \cdot \overline{t_i s}) - f(t_i)\|^2 \\
&= \sum_{i=1}^{n} \|\rho(t_i s (\overline{t_i s})^{-1}) \cdot f(\overline{t_i s}) - f(t_i)\|^2 \\
&= \sum_{i=1}^{n} \|\rho(t_i s (\overline{t_i s})^{-1}) \tfrac{v_0}{\sqrt{n}} - \tfrac{v_0}{\sqrt{n}}\|^2 \\
&= \tfrac{1}{n} \sum_{i=1}^{n} \|\rho(t_i s (\overline{t_i s})^{-1}) v_0 - v_0\|^2
\end{aligned}$$

and thus for one of the elements in Σ', say u we get
$$K(G, \Sigma) \leq \|\rho(u)v_0 - v_0\|^2.$$
Since v_0 was arbitrary we obtain as required $K(H, \Sigma') \geq K(G, \Sigma)$. $\square$

Proposition 3.10. *Let A be an abelian group generated by k elements, Σ, with $K(A, \Sigma) = \varepsilon$. Then $|A| \leq C^k$ where $C = [\frac{2\pi}{\sqrt{\varepsilon}}] + 1$.*

Proof. Divide the unit circle into C equal arcs starting at 1. This gives a subdivision on the k-torus, Π^k, into C^k boxes. Any unitary character of A determines a point of Π^k. If $|A| > C^k$ then there are two distinct characters that land in the same box and thus their quotient would be a non trivial character χ satisfying
$$|\chi(s) \cdot 1 - 1|^2 < \left(\frac{2\pi}{C}\right)^2$$

for all $s \in \Sigma$. Since χ defines a one dimensional unitary representation we conclude that
$$\varepsilon < \left(\frac{2\pi}{C}\right)^2$$
which contradicts our choice of C. Thus $|A| \leq C^k$ are required. □

Finally we can complete the proof of 3.5 as an immediate consequence of propositions 3.8 and 3.9 choosing for the constant in the theorem $([\frac{2\pi}{\sqrt{\varepsilon}}]+1)^d$.

4. $\{SL_n(p)\}$-FIXED n VERSUS FIXED p

This section is devoted to examples of finitely generated amenable groups Γ. Using theorem 3.1 this leads us to finite quotient groups which are not expanders with respect to the generators that come from Γ.

Example 4.1. Let $K = \prod_{n=5}^{\infty} S_n$, where S_n is the permutation group of $\{1, 2, \ldots, n\}$. Let $\tau = (\tau_n)$, $\sigma = (\sigma_n)$ be the elements of K defined by:
$$\tau_n = (1\,2), \text{ all } n; \quad \sigma_n = (1\ 2\ 3 \cdots n), \text{ all } n;$$
i.e. σ_n is the cyclic permutation on the n numbers $1\ 2\ 3\ \cdots n$. Let $\Gamma = \langle \sigma, \tau \rangle \subset K$ denoted the countable group generated by τ and σ.

We claim that

(i): Γ is an amenable group

(ii): if $\Gamma_+ = \Gamma \cap (\prod_{n=5}^{\infty} A_n)$, where $A_n < S_n$ is the alternating group then Γ_+ is of index 4 in Γ and is dense in $\prod_{n=5}^{\infty} A_n$.

Proof. (i) Clearly Γ is finitely generated. Let Δ_k denote the subgroup of Γ generated by $\{\tau, \sigma\tau\sigma^{-1}, \sigma^2\tau\sigma^{-2}, \ldots \sigma^{k-1}\tau\sigma^{-(k-1)}\}$. The projection of Δ_k to S_n for $n \leq k$ is surjective. On the other hand, the projection of Δ_k to $\prod_{n=k+1}^{\infty} S_n$ is a finite group, it is in fact S_k diagonally embedded in that product. Thus Δ_k is a finite group. Hence $\Delta = \bigcup_{1}^{\infty} \Delta_k$ is a locally finite group and thus is amenable.

Now $\sigma^{-1}\Delta\sigma \supset \Delta$ and thus $\tilde{\Delta} = \bigcup_{n=1}^{\infty} \sigma^{-n}\Delta\sigma^n$ is amenable as an increasing union of amenable groups and it is the normal closure of τ in Γ. Hence $\Gamma/\tilde{\Delta}$ is a cyclic group generated by the image of σ. It follows that Γ, as amenable extension of an amenable group is itself amenable.

(ii) For even n, both τ_n and σ_n are odd permutations, while for odd n, τ_n is odd and σ_n is even. Thus for a word ω in τ and σ, the sign of $(\omega)_n$ is constant along even n's and odd n's separately and the two homomorphisms
$$\text{sg}_0(\omega) = \text{ sign of the projection of } \omega \text{ into } S_n \text{ for } n \text{ even}$$
$$\text{sg}_1(\omega) = \text{ sign of projection of } \omega \text{ into } S_n \text{ for } n \text{ odd}$$
are well defined and together they define an epimorphism from Γ onto $\{\pm 1\} \times \{\pm 1\}$. Let Γ_+ denote the kernel. It is of index 4 and Γ_+ is projected onto A_n for all n. In

order to see that the image of Γ_+ is dense in $\prod_{5}^{\infty} A_n$ it suffices to show that its image in any finite product $H = \prod_{n=5}^{h} A_n$ is onto. In fact the image of Γ_+ is a subgroup L of H which is mapped onto A_n for all $5 \leq n \leq h$. Each of the simple groups A_n is a Jordan-Hölder factor of H, as well as of L. Since these groups are all different the order of L equals the order of H whence $L = H$ as required. □

P.M. Neumann called our attention that the group Γ above was already been studied for different reasons by B.H. Neumann in 1937! (see [N]).

Now theorem 3.1 implies that the family of Cayley graphs $X_n = X(S_n, \{\tau_n, \sigma_n\})$ is not a family of expanders. This is well known and easy to prove directly cf. ([L1] example 4.3.3.(c)) where three different proofs are given. The easiest one uses the fact that diameter $(X(S_n, \{\tau_n, \sigma_n\}))$ is like n^2 while the diameter of expanders grow like the logarithm of the order of the group which would be $n \log n$ in this case. We mention in passing here that the following important problem is still open:

Problem 4.2. Are there bounded sets of generators, Σ_n for S_n such that $X(S_n, \Sigma_n)$ is an expander family?

This problem is not yet solved even allowing for non constructive methods. Note that by a recent result of Alon and Roichman [AR] every finite group is an expander with respect to $O(\log |G|)$ generators- but the issue here is the bound set of generators.

Here is another example with similar flavor.

Example 4.3. Let p be a fixed prime and denote by K^p the product $\prod_{n=2}^{\infty} SL_n(p)$. Consider the embedding of $SL_2(p)$ in $SL_n(p)$ in the upper left corner. Now let Γ be the subgroup of K^p generated by $\tau = (\tau_n)$, $u = (u_n)$, $\sigma = (\sigma_n)$ where:

$$\tau_n = \begin{pmatrix} 0 & 1 \\ -1 & 0 \end{pmatrix} \in SL_2(p) \subseteq SL_n(p)$$

$$u_n = \begin{pmatrix} 1 & 1 \\ 0 & 1 \end{pmatrix} \in SL_2(p) \subseteq SL_n(p)$$

$$\sigma_n = \begin{pmatrix} 0 & 1 & \cdots & & 0 \\ \vdots & \ddots & 1 & \cdots & 0 \\ \vdots & & \ddots & \ddots & \vdots \\ \vdots & & & 0 & \ddots & 1 \\ (-1)^{n-1} & \cdots & & \cdots & 0 \end{pmatrix} \in SL_n(p).$$

We claim that Γ is a finitely generated amenable group which is dense in K^p. The proof is very similar to the one we gave for example 4.1, and we omit the details.

In [BKL] it was known that the Cayley graphs $Y_n = X(SL_n(p); \{\tau_n, u_n, \sigma_m\})$ have **logarithmic** diameter. In spite of this, using example 4.3 and theorem 3.1 we have:

Corollary 4.4. *The family of Cayley graphs $X(SL_n(p); \{\tau_n, u_n, \sigma_n\})$ is not an expander family.*

A more elementary proof of this pact was suggested to us by Yael Luz: $SL_n(p)$ acts transitively on the non zero vectors of F_p^n. If the Y_n would be expanders then their quotient graphs Z_n would be expanders, where the vertices of Z_n are $F_p^n \setminus \{0\}$ and $\alpha \in Z_n$ is adjacent to $\tau_n^{\pm 1}(\alpha)$, $u_n^{\pm 1}(\alpha)$, $\sigma_n^{\pm 1}(\alpha)$. Let $A_n \subseteq Z_n$ be the subset $\{\epsilon_3, \ldots, \epsilon_{[n/2]}\}$ where (ϵ_i) is the standard basis of F_p^n then $\tau_n^{\pm}(A_n) = u_n^{\pm}(A_n) = A_n$ while $|\sigma_n^{\pm}(A_n) \Delta A_n| \leq \frac{5}{n}|A_n|$, hence the Z_n's are not expanders.

Example 4.3 is especially interesting in light of the following:

Proposition 4.5. *For fixed n, the compact group $K_n = \prod_{p-prime} SL_n(p)$ does not contain a finitely generated dense amenable group. For $n \geq 3$, K_n contains a dense group with property (T).*

Proof. Let Γ be a finitely generated subgroup of K_n. For each $g \in \Gamma \setminus \{e\}$ there is an n-dimensional representation $\rho : \Gamma \to SL_n(p)$ for some p with $\rho(g) \neq 1$. By a lemma due to Wilson [W] (or implicitly to Malcev; see also [LMS] for explicit formulation) there exists a finite family of fields $\{K_i\}_{i \in J}$ and representations $\rho_i : \Gamma \to SL_n(K_i)$ such that $\bigcap_{i \in J} \ker \rho_i = \{e\}$. Hence Γ is imbedded in $\bigoplus_{i \in J} SL_n(K_i)$. Now if Γ is in addition amenable then by the dichotomy of Tits [T] it follows that $\rho_i(\Gamma)$ is virtually solvable for each $i \in J$ whence Γ itself is virtually solvable, i.e. has a finite index solvable subgroup. The group Γ cannot be dense in K_n since $PSL_n(p)$ are simple groups of unbounded order. This proves the first assertion of the proposition.

For the second assertion consider the diagonal embedding of $SL_n(\mathbb{Z})$ in $\prod_p SL_n(p)$. This is clearly injective and the image is dense since for every $m \in \mathbb{Z}$ the canonical projection $SL_n(\mathbb{Z}) \to SL_n(\mathbb{Z}/m\mathbb{Z})$ is onto. This is a special case of **strong approximation** which can be established here by elementary methods. □

Remarks 4.6.

(1): The group $\bigoplus_p SL_n(p)$ is an amenable group dense in K_n but is not finitely generated.

(2): K_2 does not have a dense subgroup with property T. If Γ were such a group then Γ would be separated by finitely many two dimensional representations as in the proof of proposition 4.5. But by a result of Zimmer every subgroup of $SL_2(K)$ with property T is finite (cf. [L1, thm. 3.4.7]). On the other hand K_2 does have a dense subgroup with property (τ), i.e. all **finite** representations are bounded away from the trivial one (equivalently, Γ has property (τ) if all of its finite quotients with respect to a fixed generating set are an expander family, see [L1,§4]). Such a group is $SL_2(2) \times SL_2(\mathbb{Z}[\frac{1}{2}])$, see [L1, example 4.3.3E].

(3): We do not know whether or not K^p contains a dense subgroup with property T. This appears to be quite unlikely.

The known examples of discrete groups Γ with property T are all obtained from lattices in semisimple Lie groups G which have property T. In all cases where G has property T it has been shown that all lattices Γ are arithmetic, [M3] for groups with rank ≥ 2 and [GS] for $Sp(n,1)$ and F_4. These cases will now considered separately.

(i): When Γ is a lattice in a s.s Lie group of rank ≥ 2 Serre [Se] conjectured (and his conjecture has been established in most cases) that there is an affirmative answer to the congruence subgroup problem (CSP). This in turn implies that they cannot be mapped densely into K^p.

(ii): Lattices in rank 1 Lie groups like $Sp(n,1)$ and F_4 have many infinite quotient groups [Gr] which give many more examples of exotic groups with property T. Nonetheless, the issue as to what kind of finite quotients those groups have is not so clear. Serre conjectured that such lattices should have a negative answer to the CSP. However, in light of the recent proof of super rigidity for these groups it is not unreasonable to modify his conjecture so that these lattices also satisfy the CSP. (even though they have many infinite normal groups). If this is the case then indeed also this kind of groups with property T cannot be densely embedded in $\prod_p SL_n(p)$.

5. Open problems and remarks

As we have already said in the introduction, the fundamental problem that motivated the work in this paper is whether the expander property for a family of finite groups is a property of the groups themselves or does it also depend upon the choice of generators. The contrasting results of §2 and §3 about expander and non expander families illuminate same aspects of the following open problem:

Problem 5.1. Let $\{G_i\}$ be a family of finite groups with two sets of generators $\{\Sigma_i\}$, $\{\Sigma_i'\}$ and the size of all them being bounded by a fixed constant k. Suppose that $\{G_i, \Sigma_i\}$ is an expander family, i.e. the Cayley graphs $X(G_i, \Sigma_i)$ are expanders, is also $\{G_i, \Sigma_i'\}$ an expander family.

We don't know of any family of groups that is an expander with respect to one set of generators but not with respect to another. This is rather surprising in view of the fact that in the expander families of §2 the choice of generators was crucial. In the other direction we also don't know of **even one** family of groups that is an expander family with respect to all bounded sets of generators. A natural candidate for such a family would be

$$\{SL_2(p) : p \text{ a prime}\}.$$

Computations carried out in Laffery-Rockmore [LR] suggest indeed that this might even be an expander family even if one were to choose for each p the "worst" possible generating set with only two elements. One consequence of the expander property is that the diameters of the graphs in question are logarithmic in the size of the graph. Here is weaker open question:

Problem 5.2. Is

$$\text{diameter}(X(SL_2(p), \Sigma)) \leq C \cdot \log p$$

for some fixed constant C and **all** sets of generators Σ? Is it true at least for random sets of k-generators where k is fixed and $p \to \infty$.

On the other side of the coin, we did present in §3 at least some examples of families of groups which are not expanders with respect to any sets of generators of bounded size.

A problem related to 5.1 but concerning infinite groups is as follows. If Γ is an infinite group with property T, and Σ is a finite set of generators then it is known that $K(\Gamma, \Sigma) > 0$ where $K(\Gamma, \Sigma)$ was defined in 3.4. Very little is known about exact values of $K(\Gamma, \Sigma)$ (see [Bu]) and even the following qualitative question seems to be open:

Problem 5.3. Let Γ be a group with property T. Is

$$\inf\{K(\Gamma, \Sigma) : \Sigma \text{ a set of generators for } \Gamma\}$$

strictly positive?

Since property T yields expanders while amenable groups give non expanders the following strategy, for answering problem 5.1 in the negative, suggests itself: find an infinite family of finite groups which are quotients of a finitely generated amenable group A and also a quotients of a group B with property T. For example, if K were a profinite group and A and B were finitely generated dense in K, with A amenable and B with property T then we would achieve this goal. In fact we don't believe that this is possible. Indeed proposition 4.4 above and 5.5 below lend support to the following:

Conjecture 5.4. Let K be a compact group. If A and B are both finitely generated subgroups dense in K with A amenable and B having property T then K is finite.

Proposition 5.5. *Conjecture 5.4 is valid if K is a compact Lie group, or more generally K is a linear group over some field.*

Proof. If K is a compact Lie group it is linear over $\mathbb{R}$ so that in both cases we have $K \hookrightarrow GL_n(F)$ for some field F. From the Tits alternative [T] $A \subset G$ either contains a free group on two generators or is a finite extension of a solvable group. Since A is amenable the first possibility is ruled out and thus A is virtually solvable and since it is dense in K, K is also virtually solvable which implies that B is virtually solvable. A subgroup of finite index of a group with property T has property T and if a solvable group has property T it must be finite whence B itself is finite and thus so is its closure K. □

Another approach is via invariant measures. If K, a compact group, has a dense subgroup with property T then K has an affirmative answer to the Banach-Ruziewicz problem. That means that Haar measure μ on K is the unique finitely additive measure on K, defined on μ-measurable sets, that is invariant under all group translations. (see [L1] and the references there). On the other hand, for a countable amenable group $A \subset K$ we can find many finitely additive measures on K that are invariant under A (this follows easily from the existence of approximately invariant subsets in K, cf. [Sc]). Unfortunately invariance under A doesn't imply invariance under K for finitely additive measures so this doesn't settle conjecture 5.4. In light of examples 4.1 and 4.3 the following test cases are of interest:

Problem 5.6. Let K be one of the groups:

$$\prod_{n=2}^{\infty} SL_n(p), \quad \prod_{n=1}^{\infty} S_n$$

and let μ denote Haar measure on K. Does K have a finitely additive probability measure defined on the μ-measurable sets which is different from μ.

An affirmative reply to 5.6 would imply that K doesn't contain any dense subgroups with property T. This fact alone would also follow from conjecture 5.4

References

[AB] F. Annexstein, M. Baumslag, *On the diameter and bisector size of Cayley graphs*, preprint.

[AR] N. Alon, Y. Roickman, *Random Cayley graphs and expanders*, preprint.

[BKL] L. Babai, W.M. Kantor and A. Lubotzky, *Small diameter Cayley graphs for finite simple groups*, Europ. J. of Combinatorics **10** (1989), 507–522.

[Be] H. Behr, *Finite presentability of arithmetic groups over global function fields*, Proc. Edenburgh Math. Soc. **30** (1987), 23–39.

[Bu] M. Burger, *Kazhdan constants for $SL_3(\mathbb{Z})$*, J. Reine-Angew. Math. **413** (1991), 36–67.

[C] P. Chiu, *Cubic Ramanujan graphs*, Combinatorica, to appear.

[EGM] J. Elstrodt, F. Grunewald and J. Mennicke, *Porinearé series, Kloostreman sums and eigenvalues of the Laplician for congruence groups acting on hyperbolic spaces*, C.R. Acad. Sei. Paris **305** (1987), 577–580.

[Gr] M. Gromov, *Hyperbolic groups*, in "Essays in Group Theory" (Ed: S.M. Gersten) pp. 75–264, MSRI Publications No. 8, Springer-Verlag, New York 1987.

[GS] M. Gromov, A. Schoen, in preperation.

[LR] J. Laffery, D. Rockmore, *Fast Fourier analysis for SL_2 over a finite field and related numerical experiments*, preprint.

[Li] J. Li, *Kloosterman-Selberg zeta functions on complex hyperbolic spaces*, Amer. J. of Math. **113** (1991), 653–731.

[LPSS] J. Li, I. Piatestski-Shapiro and P. Sarnak, *Pornearé series for $SO(n, 1)$*, Proc. Indian Aca. Sci (Math. Sci.) **97** (1987), 232–237.

[L1] A. Lubotzky, *Disctere Groups, Expanding Graphs and Invariant Measures*, preprint.

[L2] A. Lubotzky, *Lattices in rank one groups over local fields*, Geometric and Functional Analysis **1** (1991), 405–431.

[LMS] A. Lubotzky, A. Mann and D. Segal, *Finitely generated groups of polynomial subgroup growth*, Israel J. of Math, to appear.

[LPS] A. Lubotzky, R. Phillips and P. Sarnak, *Ramanujan graphs*, Combinatorica **8** (1988), 261–277.

[LZ] A. Lubotzky, R.J. Zimmer, *Variants of Kazhdan's property for subgroups of semi-simple groups*, Israel J. of Math. **66** (1989), 289–299.

[M1] G. Margulis, *Explicit constructions of concentrators*, Probl. of Inform. Transm. **10** (1975), 325–332.

[M2] G. Margulis, *Explicit group theoretical constructions of combinatorial schemes and their applications to the design of expanders and superconstants*, Problems of Information Transmission **24** (1988), 39–46.

[M3] G. Margulis, *Discrete Subgroups of Semi-Simple Lie Groups*, Springer, Berlin 1991.

[Mo] M. Morgenstern, *Existence and explicit constructions of $q + 1$ regular Ramanujan graphs for every prime power q*, preprint.

[MKS] W. Magnus, A. Karrass and D. Solitar, *Combinatorial Group Theory*, Wiley, New York 1966.

[N] B.H. Neumann, *Some remarks on infinite groups*, J. of the Lond. Math. Soc. **12** (1937), 120–127.

[Sc] K. Schmidt, *Amenability, Kazhdan's property T, strong ergodicity and invariant means for ergodic group actions*, Erg. Th. and Dyn. System **1** (1981), 223–236.

[S] A. Selberg, *On the estimation of Fourier coefficients of modular forms*, Proc. Symp. Pure Math. VIII (1965), 1–15.
[Se] J.P. Serre, *Le problem des groupes de congruence pour SL_2*, Ann. of Math. **92** (1970), 489–527.
[T] J. Tits, *Free subgroups in linear groups*, J. of Algebra **20** (1972), 250–270.
[W] J.S. Wilson, *Two-generator conditions for residually finite groups*, Bull. London Math. Soc. **104** (1991), 239–248.

DEPARTMENT OF MATHEMATICS, THE HEBREW UNIVERSITY, JERUSALEM, ISRAEL.
E-mail address: alexlub@@sunrise.huji.ac.il, weiss@@sunrise.huji.ac.il

RAMANUJAN GRAPHS AND DIAGRAMS FUNCTION FIELD APPROACH

MOSHE MORGENSTERN

October 1992

ABSTRACT. Let k be a local field with residue field of q elements. We investigate the quotients of the tree of $PGL_2(k)$ by some families of arithmetic lattices of $PGL_2(k)$. When these lattices are uniform this gives an explicit construction for $q+1$ regular Ramanujan graphs. In case of non-uniform lattices the quotients are more complicated, they are called diagrams. We show the close connection between the expansion properties of the diagram and the spectrum of the Laplacian on it. Then we show how to construct explicitly $q+1$ regular Ramanujan diagrams (i.e. the best possible expanding diagrams) and point to other important expanding graphs which lie in it.

0. Introduction

The technique of Lubotzky Phillips and Sarnak to explicitly construct $p+1$ regular Ramanujan graphs as quotients of the tree of $PGL_2(\mathbb{Q}_p)$ by arithmetic lattices, can also be implemented over function fields. When trying to construct $q+1$ regular Ramanujan graphs for a prime power q by passing to other number fields, many unsolved problems appear. However, over function fields this works without any difficulties. this way, in section 1 we explicitly construct many infinite families of $q+1$ regular Ramanujan Graphs.

Another essential difference when working over function fields is the appearance of non-uniform lattices. The quotients of the tree of PGL_2 by these lattices are called diagrams. These are infinite graphs, on which a finite measure is induced by the Haar measure of PGL_2. They behave therefor as finite graphs from many points of view. In section 2 we investigate the expansion properties of a diagram and its connection to the Laplacian. In section 3 we show an explicit construction for infinite Ramanujan Diagrams and deduce an explicit construction for some other very important expanding graphs called "Bounded-concentrators".

1. Uniform Lattices and Ramanujan Graphs

The technique for the explicit construction of the Lubotzky-Phillips-Sarnak Ramanujan graphs is as follows (see [Lu] , [LPS]): Let p be an odd prime number and

---

1991 *Mathematics Subject Classification.* Primary 05C35 ; Secondary 05C25 .
The final version of this paper will be submitted for publication elsewhere.

T_p is the tree of $PGL_2(\mathbb{Q}_p)$ i.e., the $p+1$ regular tree in which the vertices are the cosets of $PGL_2(\mathbb{Q}_p)/PGL_2(\mathcal{O}_p)$ (where $\mathcal{O}_p$ are the integers of $\mathbb{Q}_p$), and the $p+1$ neighbors of a vertex $gPGL_2(\mathcal{O}_p)$ are $gs_iPGL_2(\mathcal{O}_p)$ for $i=1,\ldots,p+1$, where

$$\{s_1,\cdots,s_{p+1}\} = \left\{\begin{pmatrix} p & b \\ 0 & 1 \end{pmatrix} \bigg| 0 \le b < p\right\} \cup \left\{\begin{pmatrix} 1 & 0 \\ 0 & p \end{pmatrix}\right\}.$$

Let $\Gamma \subseteq PGL_2(\mathbb{Q}_p)$ be a lattice (i.e. discrete, and there is a $PGL_2(\mathbb{Q}_p)$-invariant finite measure on $\Gamma\backslash PGL_2(\mathbb{Q}_p)$). Γ acts as a group of automorphisms of T_p (by multiplication from the left) and it is always true that at least for a finite index sublattice $\Gamma' \subseteq \Gamma$, no element $1 \ne \gamma \in \Gamma'$ fixes a vertex (or edge) of T_p. This implies that $\Gamma\backslash PGL_2(\mathbb{Q}_p)$ is compact, so Γ is uniform. Hence $\Gamma\backslash T_p$ which is compact and discrete is a finite $p+1$ regular graph.

Ramanujan conjecture (proved by Eichler and Deligne) implies that for congruence subgroups Γ of arithmetic groups of $PGL_2(\mathbb{Q}_p)$ every eigenvalue λ of $\Gamma\backslash T_p$ which is not $\pm(p+1)$, satisfies $|\lambda| \le 2\sqrt{p}$, so these graphs are Ramanujan graphs.

Lubotzky raised the question of the existence and explicit construction of k regular Ramanujan graphs for a general k. A special and natural case of this question is $k = q+1$ where q is a prime power (odd or even). Since the regularity degree of the tree of $PGL_2(\mathbb{Q}_p)$ is determined by $|\mathbb{P}^1(\mathcal{O}_p/m_p\mathcal{O}_p)| = p+1$ (where m_p is the maximal ideal of $\mathcal{O}_p$), in order to build $q+1$ regular Ramanujan graphs it is natural to replace $\mathbb{Q}_p$ by a local field whose residue filed has q elements, as it is done in [Mo 1].

Let $\mathbb{F}_q$ be the field with q elements, and $k = \mathbb{F}_q(x)$ the quotient field of $\mathbb{F}_q[x]$. k_x will denote the completion of k with respect to the x-adic valuation, $\mathcal{O}_x$ the integers and m_x its maximal ideal. Let $G = PGL_2(k_x)$, $K = PGL_2(\mathcal{O}_x)$, then $T = G/K$ is the $|\mathbb{P}^1(\mathcal{O}_x/m_x\mathcal{O}_x)| = q+1$ regular tree.

Let us first treat the case when q is odd. Let $\epsilon \in \mathbb{F}_q$ be a non-square, then

$$\mathcal{A} = k1 + ki + kj + kij \qquad i^2 = \epsilon \qquad j^2 = x - 1 \qquad ij = -ji$$

is the quaternion algebra over k which ramifies exactly at $x-1$ and $1/x$. $\mathcal{S} = \mathbb{F}_q[x]1 + \mathbb{F}_q[x]i + \mathbb{F}_q[x]j + \mathbb{F}_q[x]ij$ is a maximal order in $\mathcal{A}$ and the class number of $\mathcal{A}$ is 1. It is not hard to show that there are exactly $q+1$ solutions $\xi = 1 + cj + dij$ in $\mathcal{S}$ with $N(\xi) = \xi\bar{\xi} = \xi(1 - cj - dij) = x$. Let $\xi_1,\ldots,\xi_{\frac{q+1}{2}},\bar{\xi}_1,\ldots,\bar{\xi}_{\frac{q+1}{2}}$ be these solutions, using the unique factorization in $\mathcal{S}$ we see that

$$\Gamma(1) = \langle \xi_1,\ldots,\xi_{\frac{q+1}{2}} \mid x = 1 \rangle$$
$$= \left\{ t = a + bi + cj + dij \;\middle|\; \begin{array}{l} a-1, b \equiv 0 \bmod (x-1) \\ N(t) = a \text{ power of } x \\ x \text{ doesn't divide } t \end{array} \right\}$$

is a free group with $\frac{q+1}{2}$ generators. Let $\mathcal{A}_x = \mathcal{A} \otimes_k k_x$, since $\mathcal{A}$ splits at x there is an isomorphism $\theta : \mathcal{A}_x \to M_2(k_x)$ with $\theta(\Gamma(1)) \subseteq G = PGL_2(k_x)$. Identifying $\Gamma(1)$ with its image in G, its action on the tree T_x is simply transitive (transitive, and no $\gamma \ne 1$ fixes a vertex), so $\Gamma(1)$ can be identified with the tree T_x. Let $g(x) \in \mathbb{F}_q[x]$ be prime to $x(x-1)$, then the finite index sublattice $\Gamma(g) = \{t \in \Gamma(1) \mid t \equiv I \bmod g\}$ of $\Gamma(1)$ is also uniform, so $X_g = \Gamma(g)\backslash T_x$ is a finite $q+1$ regular graph. Drinfeld's theorem [Dr] (which is the analogue of Ramanujan's conjecture for the case of function fields) ensures that X_g is a Ramanujan graph. (For more detail see [Mo 1]).

When $g(x)$ is irreducible of even degree d, the isomorphism θ is easily expressed, and X_g can be given in a very simple explicit form: Let $\mathbb{F}_{q^d} = \mathbb{F}_q[x]/g\mathbb{F}_q[x]$ and $\underline{i} \in \mathbb{F}_{q^d}$ be a square root of ϵ. In $PGL_2(\mathbb{F}_{q^d})$ there are exactly $q+1$ matrices $\xi_1, \ldots, \xi_{q+1}$ of the form:

$$\begin{pmatrix} 1 & \gamma - \delta\underline{i} \\ (\gamma + \delta\underline{i})(x-1) & 1 \end{pmatrix} \quad \gamma, \delta \in \mathbb{F}_q, \quad \delta^2\epsilon - \gamma^2 = 1.$$

Theorem 1.1. *Assume x is not a square root (is a square root) in $\mathbb{F}_{q^d}$, then the Cayley graph of $PGL_2(\mathbb{F}_{q^d})$ (of $PSL_2(\mathbb{F}_{q^d})$) with respect to the above generators $\xi_1, \ldots, \xi_{q+1}$ is a bipartite (non-bipartite) $q+1$ regular Ramanujan graph.*

For q even and a suitable quaternion algebra, it is possible to have similar results. Again when $g(x)$ is irreducible of even degree d, it is easy to exhibit them explicitly: Let $\mathbb{F}_{q^d} = \mathbb{F}_q[x]/g\mathbb{F}_q[x]$ and $f(x) = x^2 + x + \epsilon \in \mathbb{F}_q[x]$ be irreducible (it is not hard to show the existence of such f). Let $\underline{i} \in \mathbb{F}_{q^d}$ be some root of f, then in $PGL_2(\mathbb{F}_{q^d}) = PSL_2(\mathbb{F}_{q^d})$ there are exactly $q+1$ matrices $\xi_1, \ldots, \xi_{q+1}$ of the form

$$\begin{pmatrix} 1 & \gamma + \delta\underline{i} \\ (\gamma + \delta\underline{i} + \delta)x & 1 \end{pmatrix} \quad \gamma, \delta \in \mathbb{F}_q, \quad \gamma^2 + \gamma\delta + \delta^2\epsilon = 1.$$

Theorem 1.2. *The Cayley graph of $PGL_2(\mathbb{F}_{q^d})$ with respect to the generators $\xi_1, \ldots, \xi_{q+1}$ is a non-bipartite $q+1$ regular Ramanujan graph. (For the proof and more detail, see* [Mo 1]*).*

2. Non-uniform Lattices and Diagrams

Replacing $\mathbb{Q}_p$ by $k_x = \mathbb{F}_q(x)_x$ in the last section raises the question of treating non-uniform lattices which in this case appear in $PGL_2(k_x)$. These are lattices which in their action, and the action of any of their finite index sub-lattices, on T_x is with "large" stabilizers. Hence, even though $\mu(\Gamma \backslash G) < \infty$ where μ is the Haar measure of G induced to $\Gamma \backslash G$, still $D_\Gamma = \Gamma \backslash G / K = \Gamma \backslash T_x$ is infinite. Inducing μ to $\Gamma \backslash T_x$ it becomes "atomic" and can be given explicitly as follows: If $S \subseteq \Gamma$ is the stabilizer of a vertex v (an edge e), then S is finite (it is discrete and compact), and let's define the weight of v (of e) as $w(v) = |S|^{-1}$ ($w(e) = |S|^{-1}$) then w is not more than μ induced to $\Gamma \backslash T_x$. The structure of $\Gamma \backslash T_x$ under these weights leads to the following general definition:

Definition 2.1. A *diagram* is a triple $D = (V, E, w)$, where $\Gamma = (V, E)$ is an undirected graph, $|V| \leq \aleph_0$, $w : V \cup E \to (0, 1]$ is the *weight function*, and for any $e = (u, v) \in E$, $w(e) \geq w(u), w(v)$. For $S \subseteq V$, $\mu(S) = \sum_{u \in S} w(u)$ is *the measure on D*, and we assume that $\mu(V) < \infty$. D is called *infinite* if $|V| = \aleph_0$.
Call $\theta(u, v) = w(e)/w(u)$ the *entering degree* of $e = (u, v)$ to u, and for a vertex u, $in - degree(u) = \sum_{(u,v) \in E} \theta(u, v)$. D is called *k-regular* if for every $u \in V$, $in - degree(u) = k$.

Many natural questions arise: What is an expanding diagram? What is the analogue here to the high connection between expansion and eigenvalues? (See [Ta], [AM],[Al]). These questions and more were treated in [Mo 2].

Definition 2.2. A k-regular diagram D is called an (μ_0, d)-expander if $\mu(V) = \mu_0$ and for any $S \subseteq V$ with $\mu(S) \leq \frac{\mu_0}{2}$, the set of neighbors of S, $\Gamma(S)$ satisfies $\mu(\Gamma(S) \setminus S) \geq d(1 - \mu(S)/\mu_0)\mu(S)$.
Define: $d(D) = \sup\{d \mid D \text{ is an } (\mu_0, d)-expander\}$.

Let f, g be functions on a diagram D, the inner product of f and g is defined by $\langle f, g \rangle = \int f\bar{g}d\mu = \sum_{v \in D} f(v)\overline{g(v)}w(v)$, and the Laplacian Δ in $L_2(D)$ is defined by:

$$\Delta(f)(u) = in-degree(u) \cdot f(u) - \sum_{(u,v) \in E} \theta(u,v) \cdot f(v).$$

It is easy to see that Δ is Hermitian and hence invariant on $L_2^0(D) = \{f \in L_2(D) \mid \int f d\mu = 0\}$. Let $\lambda(D) = infimum\ Spectrum(\Delta|_{L_2^0(D)})$, in [Mo 2] we prove the following three theorems:

Theorem 2.3. A k-regular connected diagram D is a $(\mu(D), \frac{4\lambda}{2\lambda+k})$ expander.

For k-regular graphs the opposite is also true i.e. $\lambda(D) \geq \frac{d^2}{4k}$. For diagrams we can prove only a weaker version of it.

Definition 2.4. Let D be a connected k regular infinite diagram. A family $\mathcal{F} = \{f_i\}_{i=1}^{\infty}$ of finite sub-diagrams (which certainly are not k regular) of D, is *essential* in D if $\lim_{i \to \infty} \mu(f_i) = \mu(D)$.
Let $d(\mathcal{F}) = \inf\{d \mid \forall f \in \mathcal{F},\ f \text{ is a } (\mu(f), d)-expander\}$ and

$$d^{ess}(D) = \sup\{d(\mathcal{F}) \mid \mathcal{F} \text{ is essential in } D\}.$$

Theorem 2.5. If D is a k-regular diagram then $\lambda(D) \geq \frac{(d^{ess}(D))^2}{4k}$.

Clearly, $d^{ess}(D) \leq d(D)$ but it is not clear if they are equal and if d^{ess} may be replaced by d in the theorem (see section 4). As for graphs, it is also true that for an infinite family $\{D_i\}_{i=1}^{\infty}$ of k-regular diagrams with $\lim_{i \to \infty} \mu(D_i) = \infty$ we have: $\limsup_{\to \infty} \lambda(D_i) \leq k - 2\sqrt{k-1}$. But for diagrams it is also possible to talk about a single one.

Theorem 2.6. For a k-regular infinite diagram D, $\lambda(D) \leq k - 2\sqrt{k-1}$.

Definition 2.7. A k-regular diagram with $\lambda(D) \geq k - 2\sqrt{k-1}$ is called a *Ramanujan diagram*.

3. Explicit Ramanujan Diagrams and Other Expanding Graphs

Let us go back to the non-uniform lattices from which the motivation to treat diagrams comes. A natural example for such a lattice in $PGL_2(k_x)$ is $\Gamma = \Gamma(1) = PGL_2(\mathbb{F}_q[1/x])$. In [Se] it is shown that the quotient $\Gamma \backslash T_x$ is the following infinite ray

$$D_\Gamma: \underset{\Lambda_0}{\bullet} \overset{e_0}{\rule{1cm}{0.4pt}} \underset{\Lambda_1}{\bullet} \overset{e_1}{\rule{1cm}{0.4pt}} \underset{\Lambda_2}{\bullet} \overset{e_2}{\rule{1cm}{0.4pt}} \underset{\Lambda_3}{\bullet} \overset{e_3}{\rule{1cm}{0.4pt}} \underset{\Lambda_4}{\bullet} \cdots$$

with weights $w(e_0)^{-1} = q(q-1)$ $w(\Lambda_0)^{-1} = q(q^2-1)$, and for $i > 0$ $w(\Lambda_i)^{-1} = w(e_i)^{-1} = (q-1)q^{i+1}$. (What is really treated in [Se] and in [Mo 3] is the lattice $PGL_2(\mathbb{F}_q[x])$ in $PGL_2(\mathbb{F}_q(x)_{1/x})$, but of course it is the same). In [Ef] it is shown by a direct computation that the spectrum of the Laplacian $\Delta|_{L_2^0(D_\Gamma)}$ is the segment $[k - 2\sqrt{k-1}, k + 2\sqrt{k-1}]$. In particular, D_Γ is a Ramanujan diagram.

Let $g \in \mathbb{F}_q[1/x]$, then $\Gamma(g) = \{\gamma \in \Gamma(1) \mid \gamma \equiv I \bmod g\}$ is a finite index sublattice of $\Gamma(1)$, so $D_{\Gamma(g)} = \Gamma(g)\backslash T_x$ is a finite cover of $D_{\Gamma(1)}$. In [Mo 3] we find the exact structure of $D_{\Gamma(g)}$ and we show that it can be given explicitly as follows: Let g be of degree d (as a polynomial in $1/x$), $\mathcal{R} = \mathbb{F}_q[1/x]/g(1/x)\mathbb{F}_q[1/x]$ be represented by all polynomials (over $\mathbb{F}_q$) of degree smaller than d, (if g is irreducible this is just $\mathbb{F}_{q^d}$). $H = PGL_2(\mathcal{R})$, $B_0 = PGL_2(\mathbb{F}_q)$,

$$B_n = \left\{ \begin{pmatrix} a & b \\ 0 & 1 \end{pmatrix} \;\middle|\; a \in \mathbb{F}_q^*, \; b \in \mathcal{R}, \; degree(b) \leq n \right\} \quad n = 1, 2, \cdots, d-1.$$

The vertices of $D_{\Gamma(g)}$ are divided into sets (levels) $L_0, L_1, \cdots, L_{d-1}$, and cusps. Where for $i = 0, 1, 2, \cdots, d-1$ $L_i = H/B_i$, edges are allowed only between L_i and L_{i+1}, and an edge exists between aB_i and bB_{i+1} iff $aB_i \cap bB_{i+1} \neq \phi$. So the edges between aB_i and bB_{i+1} are just the cosets of $B_i \cap B_{i+1}$ which are contained in $aB_i \cap bB_{i+1}$ (if there are such cosets). Everything is finite and all weights are 1. Then on every vertex a_0 of L_{d-1} an infinite cusps of the form

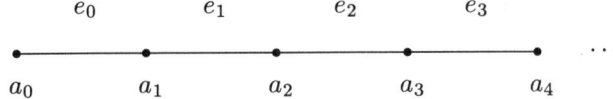

is glued, where $w(a_i) = w(e_i) = q^{-i}$.

Using representation theory of PGL_2 and Drinfeld's theorem [Dr], we prove (see [Mo 3] that $D_{\Gamma(g)}$ is a Ramanujan diagram. The only interesting part of $D_{\Gamma(g)}$ is between L_0 and L_1. It is easy to see that from L_1 on, $D_{\Gamma(g)}$ is just a "q to 1" collapsing of L_i on L_{i+1}, where from L_{d-1} on it is translated into weights which gives the cusps. For the sake of simplicity let us assume that g is irreducible, then the number of vertices in L_0 is $\frac{q^{3d}-q^d}{q(q^2-1)}$, and in L_1 it is $\frac{q^{3d}-q^d}{q^2(q^2-1)}$, so $|\frac{L_1}{L_0}| = \frac{q+1}{q}$. This gives a bipartite graph which is $q+1$ regular at L_0 and q regular at L_1. Moreover, from $\|\Delta|_{L_2^0(D_{\Gamma(g)})}\| \geq k - 2\sqrt{k-1}$ it can be concluded (see [Mo 3]) that for every $S \subseteq L_1$, $\frac{|\Gamma_0(S)|}{|S|} \geq \frac{q|L_1|}{(q-3)|S|+4|L_1|}$, where $\Gamma_0(S)$ is the set of neighbors of S which lie in L_0. From this the following theorem is immediate:

Theorem 3.1. *For $q \geq 5$ and $degree(g) \geq 2$ the subgraph of $D_{\Gamma(g)}$ which lies between L_0 and L_1 is a $(\frac{q^{3d}-q^d}{q^2(q^2-1)}, \frac{q}{q+1}, q, \frac{q-4}{q-3})$-bounded-concentrator, i.e. a bipartite graph $V = I \cup O$ with $|I| = |L_1| = \frac{q^{3d}-q^d}{q^2(q^2-1)}$, $|O| = |L_0| = \frac{q}{q+1}|I|$, it has no more than $q|I|$ edges, and every set $S \subseteq I$ with $|S| \leq \frac{q-4}{q-3}|I|$ has at least $|S|$ neighbors in O.*

These graphs are of great importance, especially for theoretical computer science. (cf. [Mo 3]).

4. Open Problems

Problem 1.
  a: Is $d^{ess}(D) = d(D)$ for every diagram D?
  b: If not, is Theorem 2.5 true, while replacing d^{ess} by d?
  c: If not, give an example of a diagram D in which $d^{ess}(D) < d(D)$, $d(D)$ is large, but $\lambda(D)$ is small.

Problem 2. For $q = 3, 4$, is the graph between L_0 and L_1 in $D_{\Gamma(g)}$ a bounded-concentrator (with some non-trivial parameters)? A positive answer for this question will give explicit bounded-concentrators which are better than what one gets by random methods!

Problem 3. In the open problems session of this conference J. Friedman asked about explicit bounded concentrators with $|O| = \sqrt{|I|}$. Here is a candidate for something similar.

For $a \in L_1$, we say that it belongs to the vertex (cusp) $b \in L_{d-1}$, if the unique path going from a to L_{d-1} through $L_2, L_3, \ldots$ (not using L_0), ends at b. Look at the following bipartite graph: $I = L_0$, $O = L_{d-1}$ and an edge exists between $a \in L_0$ and $b \in L_{d-1}$ iff for some $c \in L_1$ which belongs to b, (a, c) is an edge in $D_{\Gamma(g)}$. It is not hard to see that this graph is $q + 1$ regular in the inputs. We have reasons to believe that $D_{\Gamma(g)}$ distributes L_0 uniformly on the cusps, and that this graph is an $(\frac{q^{3d}-q^d}{q(q^2-1)}, \frac{q+1}{q^{d-1}}, q+1, \alpha)$-b.c. for some non trivial α. Is it so?

References

[Al] N. Alon, *Eigenvalues and Expanders*, Combinatorica **6** (2) (1986), 83-96.
[AM] N. Alon, V. Milman, λ_1, *Isoperimetric Inequalities for Graphs, and Superconcentrators*, J. of Combinatorial Theory, Series B **38** (1985), 73-88.
[Dr] V.G. Drinfeld, *The proof of Peterson's Conjecture for $GL(2)$ over global field of characteristic p*, Functional Analysis and its applications, **22** (1988), 28-43.
[Ef] I. Efrat, *Automorphic Spectra on the tree of PGL_2*, Publ. of M.S.R.I. no. 08908 (1986).
[LPS] A. Lubotzky, R. Phillips, P. Sarnak, *Explicit Expanders and Ramanujan Conjecture*, proc. 18th Ann. ACM Symp. on Theory of Computing 1986, 240-246.
[Lu] A. Lubotzky, *Discrete Groups, Expander Graphs and Invariant Measures*, (**preprint**).
[Mo 1] M. Morgenstern, *Existence and Explicit Construction of $q+1$ Regular Ramanujan Graphs for Every Prime Power q*, (to appear).
[Mo 2] M. Morgenstern, *Ramanujan Diagrams*, (to appear).
[Mo 3] M. Morgenstern, *Natural Bounded Concentrators*, (to appear).
[Se] J.P. Serre, *Trees*, Springer-Verlag, 1980.
[Ta] R.M. Tanner, *Explicit Concentrators from generalized N-gons*, SIAM J. of Alg. Disc. Math., **5** (1984), 287-294.

Department of Mathematics, The Hebrew University, Jerusalem, Israel.
Current address: Department of Mathematics and Computer Science, Bar-Ilan university, Ramat-Gan, Israel
E-mail address: morgen@@bimacs.cs.biu.ac.il

Highly Expanding Graphs Obtained from Dihedral Groups

HOLGER SCHELLWAT

ABSTRACT. For every odd prime p we give a construction of a bipartite p-regular Cayley graph of the dihedral group D_{p^2-1}. We show that the second largest modulus of the eigenvalues of its adjacency operator does not exceed $\sqrt{p}$.

1. Introduction

In order to minimize transmission time in a communication network it is desirable that each individual should have a large number of distinct neighbours. If one tries to meet this requirement in a trivial manner, e.g. by using a complete graph as the network topology, then one runs into a conflict with other requests, such as a (perhaps economically motivated) bound on the number of interconnecting lines. In this context the following definition proved useful.

DEFINITION 1.1. An (n, d, c)-*magnifier* is a simple graph $G = (V, E)$ on n vertices with maximal degree d such that for all subsets $X \subset V$ satisfying $|X| \leq \frac{n}{2}$ the inequality $|\partial(X)| \geq c|X|$ holds, where

$$\partial(X) = \{v \in V \setminus X : (x, v) \in E \text{ for some } x \in X\},$$

the *boundary* of X. A *family of* (d, c)-*magnifiers* is an infinite sequence of (n_i, d, c)-magnifiers, where $n_i \to \infty$ as $i \to \infty$.

Hence, the problem is to construct graphs of bounded degree with a *magnification coefficient c* as large as possible. The application problem stated above is only one paradigm for applications of magnifiers; many other construction problems in computer science can be reduced to this problem, see [9],[5],[6], and [2].

1991 *Mathematics Subject Classification*. Primary 05C25, 05C50; Secondary 20C99, 68R10.
The final version of this paper will be submitted for publication elsewhere

An $(n, d+1, c)$-expander, which is the bipartite analogue of a magnifier, can be obtained directly from an (n, d, c)-magnifier, see [1] for the details.

Although most random graphs are good expanders [13], there is a drawback in practical applications, since testing whether a graph is a superconcentrator turned out to be co–NP complete [3], and expanders [13] as well as magnifiers [14] give rise to superconcentrators. These difficulties were overcome when Alon [1] gave a spectral characterization of the magnification properties. This will be the starting-point of our construction.

2. Spectra of Cayley graphs

We restrict the scope of our investigation to finite, simple, and regular graphs on at least 3 vertices, and refer to [4] for the general terminology about graphs. Let $G = (V, E)$ be a d–regular graph, where V denotes the set of vertices, of order $|V| = n \geq 3$, and E the set of edges. The *vertex space* $L^2(V)$ of G is the linear space of complex valued functions on V and the *adjacency operator* $Q = Q(G)\colon L^2(V) \to L^2(V)$ is the linear operator on the vertex space whose matrix is the adjacency matrix. Recall the following facts from algebraic graph theory:

PROPOSITION 2.1. *Let $G = (V, E)$ be a d–regular graph, $Q = Q(G)$ its adjacency operator, and let $m(\zeta)$ denote the multiplicity of the eigenvalue $\zeta \in \mathrm{spec}(Q)$. Then Q is self-adjoint and the following conditions hold:*

(a) *If $\zeta \in \mathrm{spec}(Q)$, then $|\zeta| \leq d$*
(b) *$d \in \mathrm{spec}(Q)$*
(c) *G is connected if and only if $m(d) = 1$*
(d) *G is bipartite if and only if $-d \in \mathrm{spec}(Q)$, in this case $m(\zeta) = m(-\zeta)$ for all $\zeta \in \mathrm{spec}(Q)$.* □

If G is connected, we may define its (nontrivial) spectral radius by

$$\mu = \mu(G) := \max\{|\zeta| : \zeta \in \mathrm{spec}(Q(G)) \text{ and } |\zeta| \neq d\}.$$

Graphs with small spectral radius are good magnifiers:

PROPOSITION 2.2 (ALON [1]). *Let G be a connected d–regular graph and μ its spectral radius. Then G is an (n, d, c)–magnifier, where*

$$c \geq \frac{2d - 2\mu}{3d - 2\mu}. \quad \Box$$

Hence, the question arises how small μ can be made. There is an asymptotic limitation:

PROPOSITION 2.3 (ALON AND BOPPANA, see [11,4.2]). *Let $\{G_n\}$ be any family of d–regular graphs on n vertices, where d is fixed and n takes infinitely many values in $\mathbf{N}$. Then*

$$\liminf_{n \to \infty} \mu(G_n) \geq 2\sqrt{d-1}. \quad \Box$$

This proposition motivates the following definition.

DEFINITION 2.4. A d–regular graph G is called a *Ramanujan graph* if $\mu(G) \leq 2\sqrt{d-1}$. A *family of Ramanujan graphs* is a sequence of Ramanujan graphs on n_i vertices such that $n_i \to \infty$ as $i \to \infty$.

The only known families of Ramanujan graphs have been constructed by Lubotzky, Phillips, and Sarnak [11]. Instead of finding such families our goal is to construct graphs with spectral radius smaller than $2\sqrt{d-1}$, in fact we get $\mu(G) \leq \sqrt{d}$. If we allow the degree d to be unbounded, this does not contradict proposition 2.3. Such graphs can have better properties in real world applications, as we will see in section 4.

Now we introduce Cayley graphs.

DEFINITION 2.5. Let Γ be a finite group together with a subset S of Γ satisfying $1 \notin S$ and $S = S^{-1}$. The *Cayley graph* $G(\Gamma, S)$ of G with respect to S is the graph having the set $V = \Gamma$ of vertices and $E = \{(x,y) \in \Gamma \times \Gamma : xy^{-1} \in S\}$ of edges.

Clearly, $G(\Gamma, S)$ is regular of degree $|S|$ and connected if and only if S generates Γ.

The adjacency operator of $G(\Gamma, S)$ can be expressed in terms of the left regular representation of Γ:

PROPOSITION 2.6. *Let G be the Cayley graph of the group Γ of order $|\Gamma| = n$ with respect to the subset S, and $T \colon \Gamma \to U(n)$ its left regular representation. Then the equality*
$$Q(G(\Gamma, S)) = \sum_{s \in S} T(s)$$
holds, where the sum has to be taken in the operator algebra $\mathcal{A}(L^2(\Gamma))$.

PROOF. The left regular representation $T : \Gamma \to U(L^2(\Gamma))$ is defined by
$$T(h) \colon L^2(\Gamma) \to L^2(\Gamma),$$
$$L^2(\Gamma) \ni f \mapsto T(h)(f) \colon \Gamma \to \mathbf{C}$$
$$g \mapsto (T(h)(f))(g) = f(h^{-1}g),$$

and by the definition of Q we have
$$Qf(g) = \sum_{(g,h) \in E} f(h) = \sum_{gh^{-1} \in S} f(h) = \sum_{s \in S} f(s^{-1}g)$$
$$= \sum_{s \in S} (T(s)(f))(g)$$

for all $f \in L^2(\Gamma)$ and $g \in \Gamma$. □

The regular representation decomposes into irreducible components (see [12] for representation theory of finite groups), inducing a direct sum decomposition of the adjacency operator:

PROPOSITION 2.7. *Let* $\Gamma, S,$ *and* Q *be as above, and let* $T_1, \ldots, T_m$ *be a complete set of irreducible representations* $T_k \colon \Gamma \to U(n_k)$ *of* $\Gamma,$ *where* $1 \leq k \leq m.$ *Then the following conditions hold:*

(a) Q *is unitarily equivalent to*

$$\sum_{s \in S} \bigoplus_{k=1}^{m} \bigoplus_{i=1}^{n_k} T_k(s) = \bigoplus_{k=1}^{m} \bigoplus_{i=1}^{n_k} \sum_{s \in S} T_k(s)$$

(b) $G(\Gamma, S)$ *is connected if and only if*

$$|S| \in \operatorname{spec}\left(\sum_{s \in S} T_k(s)\right) \Leftrightarrow T_k \text{ is trivial}$$

(c) *If* $G(\Gamma, S)$ *is connected, then*

$$\mu(G(\Gamma, S)) \leq \max\{\mu_k : 1 \leq k \leq m\},$$

where $\mu_k := \max\left(\{|\alpha| : |\alpha| \neq |S| \wedge \alpha \in \operatorname{spec}\left(\sum_{s \in S} T_k(s)\right)\} \cup \{0\}\right).$

PROOF. The regular representation $T \colon \Gamma \to U(n)$ decomposes into irreducible components

$$T = \bigoplus_{k=1}^{m} n_k T_k = \bigoplus_{k=1}^{m} \bigoplus_{i=1}^{n_k} T_k,$$

where $n_k = \dim(T_k)$ denotes the degree of the representation T_k. Since summation over S in the algebra $\mathcal{A}(L^2(\Gamma))$ does not affect the direct sum decomposition, proposition 2.6 proves part (a).

Note that all eigenvalues of Q occur among the eigenvalues of the components $Q_k := \sum_{s \in S} T_k(s)$. If $m_k(\zeta)$ denotes the multiplicity of the eigenvalue $\zeta \in \operatorname{spec}(Q_k)$, then its multiplicity $m(\zeta)$ in $\operatorname{spec}(Q)$ is given by $n_k \cdot m_k(\zeta)$. The degree, and hence the multiplicity of the trivial representation which yields the eigenvalue $|S|$ in any complete set of irreducible representations equals 1. This observation completes the proof of (b) and (c) by proposition 2.1. □

In the next section we will use these results to determine the spectral radius of Cayley graphs of dihedral groups.

3. Cayley graphs of dihedral groups

The structure of a dihedral group resembles that of a cyclic group as it contains a cyclic subgroup of index 2. So difficulties have to be expected in the construction of magnifiers from dihedral groups, since cyclic groups do not make good magnifiers (see [9]) if one chooses the standard generators for S. Hence, the generators of the Cayley graph have to be chosen judiciously in order to yield a small spectral radius. This can be accomplished using the following theorem:

THEOREM 3.1 (KATZ [8]). *Let π be a nontrivial complex-valued multiplicative character defined on an extension field $\mathbf{E}$ over a finite field $\mathbf{F}$ with dimension $[\mathbf{E}\colon \mathbf{F}] = t$ and $\omega \in \mathbf{E}$ any element that generates $\mathbf{E}$ over $\mathbf{F}$. Then*

$$\left|\sum_{f\in\mathbf{F}} \pi(f+\omega)\right| \leq (t-1)\sqrt{|\mathbf{F}|}. \quad \square$$

Lubotzky [10] and Chung [6] used theorem 3.1 to construct Ramanujan graphs from cyclic groups. Our construction is similar, yet it leaves no uncertainty about the cardinality of S. For every odd prime p we will construct a Cayley graph of the dihedral group $\mathbf{D}_{p^2-1}$. In order to apply proposition 2.7 we need a complete set of irreducible representations of the group $\mathbf{D}_n$ for $n = p^2 - 1$. Such a set can be found in [15], we just state the results.

The group $\mathbf{D}_n$ can be presented as $\mathbf{D}_n = \{r, s; r^n = s^2 = (sr)^2 = 1\}$. Every $\alpha \in \mathbf{D}_n$ can be written either as $\alpha = r^k$ or $\alpha = sr^k$, with a suitable $k \in \mathbf{N}$. First, there are 4 representations of degree 1, hence they coincide with their characters which are given in the following table:

	r^k	sr^k
ψ_1	1	1
ψ_2	1	-1
ψ_3	$(-1)^k$	$(-1)^k$
ψ_4	$(-1)^k$	$(-1)^{k+1}$

These are completed by $\frac{n}{2} - 1$ pairwise inequivalent irreducible representations ρ_h of degree 2. Let h be an integer satisfying $0 < h < n/2$ and $\theta := \exp(2\pi i/n)$. Then ρ_h is given by

$$\rho_h(r^k) = \begin{pmatrix} \theta^{hk} & 0 \\ 0 & \theta^{-hk} \end{pmatrix}, \quad \rho_h(sr^k) = \begin{pmatrix} 0 & \theta^{-hk} \\ \theta^{hk} & 0 \end{pmatrix}.$$

Now we give the construction of the Cayley graph.

PROPOSITION 3.2. *For every odd prime p there is a subset $S \subset \mathbf{D}_{p^2-1}$ such that the Cayley graph $G = G(\mathbf{D}_{p^2-1}, S)$ is a connected, bipartite, and p-regular graph on $2p^2 - 2$ vertices whose spectral radius $\mu(G)$ does not exceed $\sqrt{p}$.*

PROOF. Clearly, the graph has $2p^2 - 2$ vertices, since the order of $\mathbf{D}_n$ is $2n$ and $n = p^2 - 1$. Let $\mathbf{E}$ denote the quadratic extension of $\mathbf{F} := \mathbf{F}_p$ and choose a generator g of the group $\mathbf{E}^*$ of invertible elements of the finite field $\mathbf{E} = \mathbf{F}_{p^2}$ and an element ω that generates $\mathbf{E}$ over $\mathbf{F}$. Then we have $[\mathbf{E}\colon \mathbf{F}] = 2$, $\mathbf{E} = \mathbf{F}(\omega)$, $\mathbf{E}^* = \langle g \rangle$, and $\mathbf{E}^*$ can be identified with the cyclic subgroup $U := \langle r \rangle$ of $\mathbf{D}_n$. Put $\overline{S} := \{\omega + f : f \in \mathbf{F}\}$. Since $\langle g \rangle = \mathbf{E}^*$ and $0 \notin S$ every element $\omega + f \in \overline{S}$ can be written as $\omega + f = g^{a_f}$ for some uniquely determined integer $a_f \in \{1, \ldots p^2-1\}$. These integers define our generating set for the Cayley graph by $S := \{sr^{a_f} : f \in \mathbf{F}\}$. This is an admissible set in the sense of definition 2.5.

Indeed, $1 \notin S$ because $\overline{S} \subseteq \mathbf{E}\setminus\mathbf{F}$ and $S = S^{-1}$ because $(sr^k)^2 = 1$. Furthermore, the degree of G is $|S| = |\mathbf{F}| = p$.

Now we apply proposition 2.7 to determine the spectrum of $Q(G)$. We have to consider the operators $Q_k = \sum_{s \in S} T_k(s)$ given by their associate matrices $A_k = \sum_{s \in S} A_k(s)$. First, we check the representations of degree 1:

(a) $A_1 := \sum_{s \in S} \psi_1(s) = \sum_{s \in S} 1 = |S| = p \in \text{spec}(Q(G))$. In particular, the multiplicity of p is 1.

(b) $A_2 := \sum_{s \in S} \psi_2(s) = \sum_{s \in S} -1 = -|S| = -p \in \text{spec}(Q(G))$, hence G is bipartite by proposition 2.1.

(c) $A_3 := \sum_{s \in S} \psi_3(s) = \sum_{f \in \mathbf{F}} \psi_3(sr^{a_f}) = \sum_{f \in \mathbf{F}} (-1)^{a_f}$. Now $\pi_3 \colon \mathbf{E}^* \to U(1)$, $g^k \mapsto (-1)^k$ is a nontrivial multiplicative character of $\mathbf{E}$ with $\pi_3(\omega + f) = \pi_3(g^{a_f}) = (-1)^{a_f}$ for $f \in \mathbf{F}$. Hence,

$$A_3 = \sum_{f \in \mathbf{F}} (-1)^{a_f} = \sum_{f \in \mathbf{F}} \pi_3(\omega + f)$$

and by Theorem 3.1 we have

$$|A_3| = \left|\sum_{f \in \mathbf{F}} \pi_3(\omega + f)\right| \leq (2-1)\sqrt{|\mathbf{F}|} = \sqrt{p}.$$

(d) $A_4 := \sum_{s \in S} \psi_4(s) = -\sum_{s \in S} \psi_3(s) = -A_3$. Again, $|A_4| \leq \sqrt{p}$ holds.

Now we consider the representations of degree 2. Their associate matrices are

$$A_h := \sum_{s \in S} \rho_h(s) = \sum_{f \in \mathbf{F}} \rho_h(sr^{a_f}) = \sum_{f \in \mathbf{F}} \begin{pmatrix} 0 & \theta^{-ha_f} \\ \theta^{ha_f} & 0 \end{pmatrix},$$

where $0 < h < \frac{n}{2}$. By putting $\alpha_h := \sum_{f \in \mathbf{F}} \theta^{ha_f}$ we find

$$A_h = \begin{pmatrix} 0 & \overline{\alpha_h} \\ \alpha_h & 0 \end{pmatrix}.$$

Define the characters $\pi_h \colon \mathbf{E}^* \to U(1)$ by $g^k \mapsto \theta^{hk}$. These are nontrivial multiplicative characters satisfying $\pi_h(\omega + f) = \pi_h(g^{a_f}) = \theta^{ha_f}$ for $f \in \mathbf{F}$, hence $\alpha_h = \sum_{f \in \mathbf{F}} \pi_h(\omega + f)$. Theorem 3.1 yields $|\alpha_h| \leq \sqrt{p}$, and since $\text{spec}(A_h) = \{|\alpha_h|, -|\alpha_h|\}$ the inequality $|\zeta| \leq \sqrt{p}$ holds for all $\zeta \in \text{spec}(A_h)$. Altogether we conclude that the multiplicities of the eigenvalues p and $-p$ in $\text{spec}(Q)$ are 1, showing that G is connected and bipartite, and that the spectral radius of G is bounded by $\sqrt{p}$. □

4. Applications

Let us illustrate the results of section 3 by comparing some properties of the graphs constructed there with those of Ramanujan graphs. Let G_p denote the Cayley graph of the $\mathbf{D}_{p^2-1}$ and G'_p a p-regular Ramanujan graph of order

$2p^2 - 2$ and spectral radius $2\sqrt{p-1}$ with magnification coefficients $c(G_p)$ and $c(G'_p)$, respectively. By proposition 2.2 we have

$$c(G_p) \geq \frac{2p - 2\sqrt{p}}{3p - 2\sqrt{p}},$$

which is substantially better than

$$c(G'_p) \geq \frac{2p - 4\sqrt{p-1}}{3p - 4\sqrt{p-1}}.$$

Another desirable property of communication networks is high fault tolerance. This can be expressed by a high vertex connectivity of the graph G, which we denote by $\kappa(G)$. By a result of Fiedler [7] the vertex connectivity of a d-regular graph G can be bounded by $\kappa(G) \geq d - \mu(G)$. Hence, we have $\kappa(G_p) \geq p - \sqrt{p}$, which is much better than $\kappa(G'_p) \geq p - 2\sqrt{p-1}$ for Ramanujan graphs.

References

1. N. Alon, *Eigenvalues and expanders*, Combinatorica **6** (1986), 83–96.
2. F. Bien, *Constructions of telephone networks by group representations*, Notices A.M.S. **36 (1)** (1989), 5–22.
3. M. Blum, R. Karp, O. Vornberger, C. Papadimitriou and M. Yannakakis, *The complexity of testing whether a graph is a superconcentrator*, Inform. Process. Letters **13** (1981), 164–167.
4. B. Bollobás, *Graph Theory*, Springer Verlag, New York, 1979.
5. F. R. K. Chung, *On Concentrators, superconcentrators, and nonblocking networks*, Bell System Tech. **58** (1979), 1765-1777.
6. _____, *Diameters and Eigenvalues*, J. of the Amer. Math. Soc. **2** (1989), 187–196.
7. M. Fiedler, *An algebraic approach to connectivity of graphs*, Recent Advances in Graph Theory (M. Fiedler, ed.), Academia Praha, 1975, pp. 193–196.
8. N. M. Katz, *An estimate for character sums*, J. of the Amer. Math. Soc. **2** (1989), 197–200.
9. M. Klawe, *Limitations on explicit constructions of expanding graphs*, SIAM J. Comput. **13** (1984), 156–166.
10. A. Lubotzky, *Discrete groups, expanding graphs and invariant measures* (to appear).
11. A. Lubotzky, R. Phillips, and P. Sarnak, *Ramanujan graphs*, Combinatorica **8 (3)** (1988), 261–277.
12. M. A. Naimark and A. I. Štern, *Theory of group representations*, Springer Verlag, New York - Heidelberg - Berlin, 1982.
13. N. Pippenger, *Superconcentrators*, SIAM J. Comput. **6** (1977), 298-304.
14. H. Schellwat, *Explicit undirected superconcentrators of density 21*, in preparation.
15. J. P. Serre, *Linear Representations of Finite Groups*, Springer-Verlag, New York, 1977.

Technische Universität Braunschweig, Institut für Analysis, Pockelsstrasse 14, W-3300 Braunschweig, Germany

Current address: Am Heuer 6, W-3420 Herzberg, Germany

E-mail address: I1010902@dbstu1.bitnet

Are Finite Upper Half Plane Graphs Ramanujan?

Audrey Terras

Ramanujan graphs (defined in (2) below) first appeared in Lubotsky, Phillips and Sarnak [24]. See also Fan Chung [14]. [15], Lubotsky [23] and, of course, the present volume. Such graphs are of interest for many sorts of applications in telephone networks and computer science because they are good expander graphs (defined just before (3) below). The Ramanujan graphs are of interest to number theorists because their Ihara zeta functions satisfy the Riemann hypothesis (see Sunada [39]). And they are connected with the estimation of some interesting exponential sums.

In several papers, Terras [40], Celniker et al [13], Angel et al [3], [4], we have considered some graphs associated to finite analogs of the Poincaré upper half plane. We also found that many of the results on real symmetric spaces discussed in Terras [41] have finite analogs; e.g., there are analogs of K-Bessel functions, spherical functions, the uncertainty principle, Selberg's trace formula. Here we shall just summarize some of this work, centering on the analysis that ultimately led N. Katz [21] to estimate the last remaining exponential sums needed to show that our graphs are indeed Ramanujan.

The work that we are summarizing is joint with Jeff Angel, Nancy Celniker, Steve Poulos, Cindy Trimble, and Elinor Velasquez. We must also thank Amy Avakian, Carol Chang, Ron Evans, Perla Myers, and Harold Stark for many helpful discussions.

The graphs to be considered are attached to a finite symmetric space which we can view as an analog of the Poincaré upper half plane (see (9)) or as G/K where $G = GL(2,\mathbb{F}_q)$ is the

---

1991 *Mathematics Subject Classification.* 11T23, 11E57.
Research partially supported by M.S.R.I., Berkeley.
Preliminary Announcement.

general linear group of invertible 2x2 matrices with entries in the field $\mathbb{F}_q$ with $q=p^r$ elements. Here p is assumed to be an odd prime. The subgroup K is

$$(1) \qquad K = \left\{ \begin{bmatrix} a & b\delta \\ b & a \end{bmatrix} \;\middle|\; a,b \in \mathbb{F}_q, \; a^2 - b^2\delta \neq 0 \right\},$$

where δ is a fixed non-square in $\mathbb{F}_q$. This is our analog of the orthogonal group and G/K is an analog of the Poincaré upper half plane. It is a finite symmetric space in the sense that the algebra $L^2(K\backslash G/K)$, consisting of all K-bi-invariant functions on G, is a commutative algebra under convolution on G. We will associate $q-2$ graphs with G/K, each graph having $q(q-1)$ vertices and degree $q+1$.

We are interested in the question of whether our graphs satisfy the following definition.

Definition. A k-regular graph is <u>Ramanujan</u> if for all eigenvalues λ of the adjacency matrix with $|\lambda| \neq k$, we have

$$(2) \qquad |\lambda| \leq 2\sqrt{k-1}.$$

Such graphs are of possible use in building communications networks because they have large expansion constants.

Definition. A finite k-regular graph X with n vertices has <u>expansion</u> <u>constant</u> c if for every set $A \subset X$ with $|A| \leq \frac{n}{2}$, defining the boundary of A as

$$\partial A = \left\{ b \in X-A \;\middle|\; b \text{ is adjacent to some } a \in A \right\}$$

we have $|\partial A| \geq c|A|$. It can be shown that if λ_1 is the second largest eigenvalue of the adjacency matrix of the graph X, then

$$(3) \qquad c = \frac{k - \lambda_1}{2k}.$$

See Sarnak [31, p. 69]. Thus if λ_1 is small, the expansion constant c is large. Some references are: Bien [6], Buck[9], Chung [14,15], Friedman [17], Lubotsky [23], Sarnak [31] for more information on Ramanujan graphs.

The name "Ramanujan" got attached to these graphs because the Ramanujan conjecture bounding Fourier coefficients of holomorphic modular forms is needed to show that the graphs of Lubotsky, Phillips and Sarnak [24] are Ramanujan. Thus, the proof requires Deligne's proof of the Ramanujan conjecture.

This last remark leads one to suspect that the work on Ramanujan graphs has connections with many of the golden threads running through the fabric of number theory; e.g., the estimation of exponential sums, the number of points on curves over finite fields, the Weil conjectures.

Before discussing these connections, we need to return to our main subject - the (zonal) spherical function - a K-invariant eigenfunction of all the G-invariant differential operators on G/K, normalized to have the value 1 on the coset K (the origin of the symmetric space G/K). Each G-invariant operator will be the adjacency operator of one of our graphs on G/K.

And $L^2(K \backslash G/K)$ is a commutative algebra of K-bi-invariant functions on G under the operation of convolution over G. So we have a symmetric space G/K - or a finite Gelfand pair for $G = GL(2, \mathbb{F}_q)$ and K as in (1). See Diaconis [16] and Terras [42] for more information on finite symmetric spaces. Another reference is Lusztig [25]. It turns out that spherical functions are both eigenfunction and eigenvalue. See equation (13). Thus the question of whether our graphs are Ramanujan reduces to an estimate for a spherical function for G/K.

By using group representation theory, one can identify the two types of spherical functions for $G = GL(2, \mathbb{F}_q)$ and K as in (1). This was done by Soto-Andrade [36]. For the principal series representations of G, the eigenvalues are

$$(4) \qquad \lambda_1(a,b) = \sum_{\substack{x^2 = ay + \delta(y-1)^2 \\ y = \delta^e}} exp(2\pi i b e/(q-1)) .$$

Here we take δ to be a generator of the multiplicative group $\mathbb{F}_q^*$. The sum in (4) is over all pairs (x,y) in $\mathbb{F}_q \times \mathbb{F}_q^*$ solving the given equation. The parameter a fixes the graph. Here a is not equal to 0 or 4δ, in order for the graph to be $q+1$ regular. We called the eigenvalues (4) one-dimensional in Angel et al [3] because they correspond to one-dimensional representations of the affine group of matrices $\begin{pmatrix} y & x \\ 0 & 1 \end{pmatrix}$ in $G = GL(2, \mathbb{F}_q)$.

The discrete series representations of G yield eigenvalues

given in terms of $c = \frac{a}{\delta} - 2$, ε a character of $\mathbb{F}_q$ which is 1 on squares, -1 on non-squares and 0 on 0, and ω a multiplicative character of $\mathbb{F}_q(\sqrt{\delta})^*$ such that $\omega \neq \omega^q$:

$$(5) \quad \lambda_s(c,\omega) = \sum_{Nz=1,\ z=x+y\sqrt{\delta}} \varepsilon(c+2x)\,\omega(z) ,$$

where the sum is over all z in $\mathbb{F}_q(\sqrt{\delta})$ with $Nz = z\bar{z} = z^{1+q} = 1$. We call these eigenvalues <u>Soto-Andrade eigenvalues</u> since he worked out this formula in [36].

The estimation of exponential sums such as (4) is connected with the Riemann hypothesis for the zeta function of a function field over a finite field. This is A. Weil's famous theorem [47]. The proof has been simplified. See Schmidt [34]. R. Evans and H. Stark both showed independently that

$$(6) \quad |\lambda_1(a,b)| \leq 2\sqrt{q}, \quad a \neq 0, 4\delta, \quad 1 \leq b \leq q-1 ,$$

using Schmidt [34, Theorem 2C, p. 43].

N. Katz [21] has showed that

$$(7) \quad |\lambda_s(c,\omega)| \leq 2\sqrt{q}, \quad \omega \neq \omega^q,\ 0 \leq c \leq q-1,\ c \neq \pm 2 .$$

The proof requires the Grothendieck-Lefschetz trace formula to see that the Soto-Andrade sum (5) is an alternating sum of traces of the Frobenius acting on certain l-adic étale cohomology groups. this reduces the proof of (7) to Weil's proof of the Riemann hypothesis for curves over finite fields.

Some references for the estimation of exponential sums are Adolphson [1], Katz [20], Serre [35], Weil [47, Vol. I, pp. 387-389, 399-410].

We can think of the spherical functions (4) and (5) as finite analogs of eigenfunctions of the non-Euclidean Laplacian on the Poincaré upper half plane. Figure 1 can be viewed as a density plot of a finite spherical function, since these functions are constant on the K-orbits in G/K. Thus Figure 1 is a finite analog of the density plots of square integrable $SL(2,\mathbb{Z})$-invariant eigenfunctions of the non-Euclidean Laplacian on the Poincaré upper half plane which are given by Hejhal and Rackner [19]. See also Sarnak [33].

In connection with quantum chaos, physicists have already been considering physical systems over other fields of numbers than ℝ. See Thiran et al [43] for a discussion of p-adic dynamical systems. Martin et al [26], and Nambu [28] consider dynamical systems over finite fields. In this context, a discussion of finite analogs of symmetric spaces seems quite natural.

So we will replace the real numbers ℝ with the finite field $\mathbb{F}_q$ having $q=p^r$ elements, for an odd prime p. And we replace the complex numbers $\mathbb{C} = \mathbb{R}[\sqrt{-1}]$ with $\mathbb{F}_q[\sqrt{\delta}]$, for δ a non-square in $\mathbb{F}_q$. Usually we take δ to be a generator of the multiplicative group of $\mathbb{F}_q$. We now give a brief summary of the basic facts about finite upper half planes and the graphs associated with them. See Angel [2], Angel et al [3], [4], Celniker [12], Celniker et al [13], Poulos [30], Terras [40], [42], Velasquez [45], [46] for more details.

Definition. The <u>finite upper half plane</u> H_q is defined by

(9) $$H_q = \left\{ z=x+y\sqrt{\delta} \;\middle|\; x,y \in \mathbb{F}_q,\; y \neq 0 \right\}.$$

Since we have no analog of positive y in a finite field, it is easier to say $y \neq 0$. So we are really looking at a double covering of an upper half plane; i.e., a union of an "upper" and a "lower" half plane.

Notation. For $z=x+y\sqrt{\delta}$, with $x,y \in \mathbb{F}_q$, write

$$x = \text{Re } z, \quad y = \text{Im } z, \quad \bar{z} = x-y\sqrt{\delta} = z^q, \quad Nz = z\bar{z}.$$

It is easy to see that $G = GL(2,\mathbb{F}_q)$ acts transitively on H_q by <u>fractional linear transformation</u>

$$\begin{pmatrix} a & b \\ c & d \end{pmatrix} z = \frac{az+b}{cz+d},$$

since $z=x+y\sqrt{\delta} = \begin{pmatrix} y & x \\ 0 & 1 \end{pmatrix} \sqrt{\delta}$. A simple calculation shows that the subgroup K defined in (1) is the <u>isotropy subgroup</u> consisting of the $g \in G$ fixing $\sqrt{\delta}$. This means that we can identify H_q with G/K.

Definition. The "<u>distance</u>" on H_q is defined by

(10) $$d(z,w) = \frac{N(z-w)}{Imz\ Imw}, \text{ for } z,w \in H_q.$$

This is not a metric since it has values in the finite field $\mathbb{F}_q$ rather than $\mathbb{R}$, but it is a natural analog of the Poincaré metric on the real upper half plane, as well as the distance on the p-adic upper half plane defined by Stark [38], since
$$d(gz,gw) = d(z,w), \text{ for all } g \in G.$$

Definition. Fix δ to be a generator of the multiplicative group $\mathbb{F}_q^*$ and let $a \in \mathbb{F}_q$. Define the <u>finite upper half plane graph</u> $X_q(\delta,a)$ to have as vertices the points of H_q. Connect vertex z to vertex w if $d(z,w)=a$.

When $a \neq 0$ or 4δ, it can be shown that $X_q(\delta,a)$ is a $(q+1)$-regular graph. The points $z=x+y\sqrt{\delta}$ adjacent to $\sqrt{\delta}$ are those with $x,y \in \mathbb{F}_q$ such that $y \neq 0$ and

(11) $$x^2 = ay + \delta(y-1)^2.$$

The graph $X_3(-1,1)$ is the octahedron. H. Stark noticed that the three graphs for $q=5$ can be placed on the dodecahedron. For example, $X_5(2,1)$ puts a star on each pentagonal face of the dodecahedron. See Terras [42] and Celniker et al [13] for the pictures. These graphs are connected with fundamental domains for congruence subgroups of the modular group $SL(2,\mathbb{Z})$ (see Angel et al [3]). More information on these graphs can be found in Celniker [12] and Poulos [30]. Angel [2] covers the case of even q. When $q=4$, he finds that the graph is the icosahedron.

Consider the sets

(12) $$S_a = S_q(\delta,a) = \left\{ x+y\sqrt{\delta} \ \middle| \ x^2=ay+\delta(y-1)^2, x,y \in \mathbb{F}_q, y \neq 0 \right\}.$$

Note that S_a consists of all $z \in H_q$ such that $d(z,\sqrt{\delta})=a$. Poulos [30] shows that these sets S_a represent the K-orbits in H_q; i.e., the K-double cosets in G, for $G = GL(2,\mathbb{F}_q)$ and K as in (1). So the adjacency operators associated to the S_a (including those for $a=0$ and 4δ) are the G-invariant operators on $L^2(H_q)$; i.e., the finite analogs of Laplacians. They may also be viewed as convolution with δ_{S_a} - delta functions of the sets S_a. This means that the algebra of these G-invariant operators on H_q is the algebra $L^2(K\backslash G/K)$ of K-bi-invariant functions on G under convolution on G. Since the sets S_a are closed under the operation

$$\begin{bmatrix} y & x \\ 0 & 1 \end{bmatrix} \mapsto \begin{bmatrix} y & x \\ 0 & 1 \end{bmatrix}^{-1},$$

the algebra $L^2(K\backslash G/K)$ is commutative and G/K is a symmetric space. We also call (G,K) a Gelfand pair. See Terras [42] for a discussion of the equivalence of this with other criteria for a symmetric space such as that of Selberg.

Many people (going back to Laplace and Legendre) have studied spherical functions on symmetric spaces. For compact real symmetric spaces G/K, Cartan (1929) viewed the spherical functions as matrix entries of the irreducible representations π of G such that $\pi(K)$ has a fixed vector. See Terras [41, Vol. II, Section 4.2.3].

The theory of spherical functions for finite symmetric spaces is rather recent. Once case of interest in coding theory is that of the Krawtchouk polynomials, which play a large role in the theory of perfect codes. See Bannai and Ito [5], Diaconis [16], Stanton [37], Terras [42], Velasquez [46]. Much of the interest has been in the symmetric space G/B, for the Borel subgroup B. If $G = GL(2,\mathbb{F}_q)$, then B is the subgroup of upper triangular matrices in G. The space G/B is analogous to the boundary of the Poincaré upper half plane. Here we shall only look at G/K, with K as in (1), $G = GL(2,\mathbb{F}_q)$.

There are three equivalent definitions of spherical functions on a finite symmetric space. We should perhaps call these functions zonal spherical functions, but we will drop the word "zonal". The following Theorem is proved in [42].

Theorem. 3 Equivalent Definitions of Spherical Functions on G/K.
A K-bi-invariant function $h:G \longrightarrow \mathbb{C}$ is spherical if it satisfies any of the following equivalent criteria.

1) The K bi-invariant function h such that $h(e)=1$, $e=$the identity of G, is an eigenfunction of all the convolution operators by f in $L^2(K\backslash G/K)$; i.e.,

$$(f * h)(x) = \lambda_f h(x), \text{ for all } f \in L^2(K\backslash G/K).$$

2) The mapping of $f \in L^2(K\backslash G/K)$ to $(f*h)(e)$ yields an algebra homomorphism of $L^2(K\backslash G/K)$ to $\mathbb{C}$.

3) The function h is an eigenfunction of the mean-value

operators

$$\text{(13)} \qquad \frac{1}{|K|} \sum_{k \in K} h(xky) = h(x)h(y), \quad \text{for all } x,y \in G.$$

Note that (13) says that the spherical function is both eigenfunction and eigenvalue of the convolution operators.

In order to get a feeling for the chaos inherent in the spherical functions on H_q, note that they are constant on the sets S_a of (12) which are the K-orbits in H_q. Thus we get an idea of the "level curves" of the spherical functions by graphing these double cosets in H_q.

In Figure 1, we show H_q as the set of (x,y) with

$$-\frac{p-1}{2} \le x \le \frac{p-1}{2} \quad \text{and} \quad 1 \le y \le p-1.$$

Here $p=163$. We color the square at (x,y) according to the value of the distance

$$\text{(14)} \qquad f(x,y) = d(x+y\sqrt{\delta},\sqrt{\delta}) = \frac{x^2 - \delta(y-1)^2}{y}.$$

Recall that $f(x,y)=a$ means that $x+y\sqrt{\delta}$ is in the set S_a, the set of neighbors of $\sqrt{\delta}$ in the graph $X_q(\delta,a)$. The figure was produced by MATHEMATICA.

Question. What can you see in such Figures?
Answers.

- symmetry under $x \mapsto -x$,
- chaos as $p \longrightarrow \infty$,
- "circles" just as in Hejhal and Rackner [19].

FINITE UPPER HALF PLANE GRAPHS

Figure 1. ListDensityPlot*[m]*, *p=163*.

m = the matrix of values of the distance $f(x,y)=d(x+y\sqrt{\delta},\sqrt{\delta})$, for $p=163$, $-\frac{p-1}{2} \leq x \leq \frac{p-1}{2}$, $1 \leq y \leq p-1$. The distance $d(z,w)$ was defined in (10). The plot was produced by MATHEMATICA.

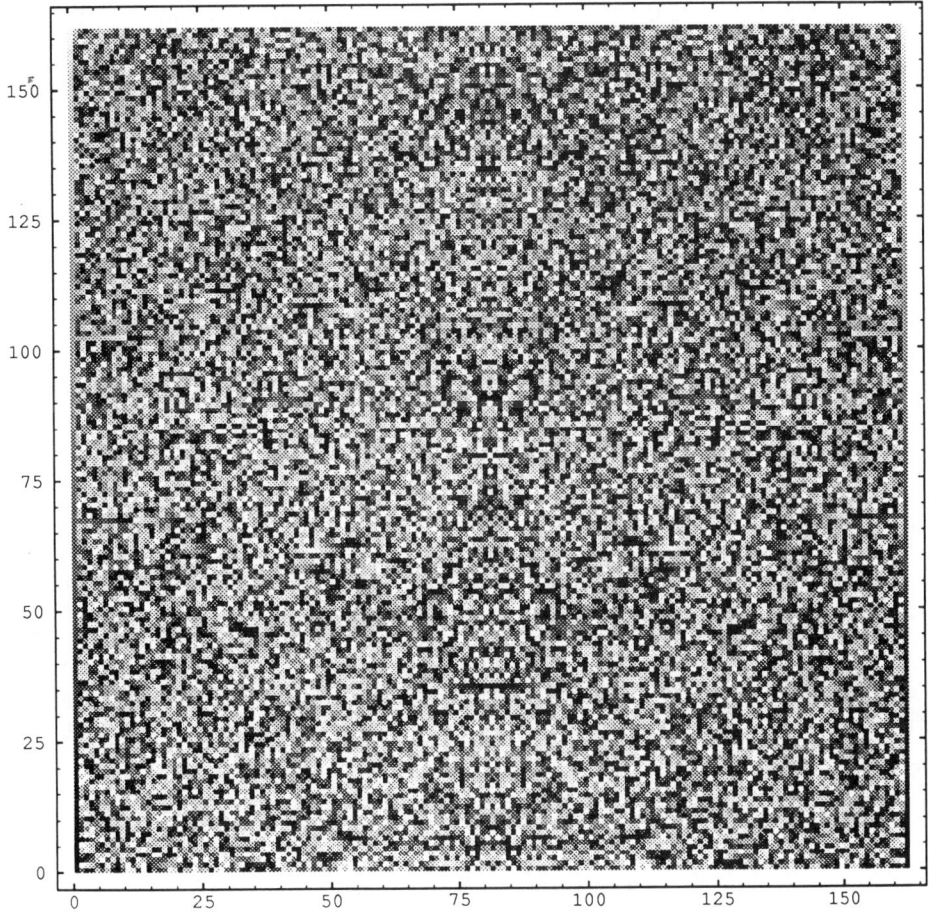

Suppose that $\hat{G}$ denotes a complete set of irreducible unitary representations π of G; i.e., π is a group homomorphism of G into $U(d_\pi)$, the group of unitary $d_\pi \times d_\pi$ complex matrices. Here "irreducible" means not uniformly block diagonalizable. See Fulton and Harris [18] or Piatetski-Shapiro [29] for descriptions of the irreducible unitary representations of $GL(2,\mathbb{F}_q)$. There are two main types.

The Two Main Types of Representations of $G = GL(2,\mathbb{F}_q)$.

I. **Principal Series.** These representations are induced from a representation ρ of B the Borel subgroup of upper triangular matrices in G. Here ρ comes from two multiplicative characters χ, κ of $\mathbb{F}_q^*$ via

$$\rho \begin{bmatrix} a & b \\ 0 & c \end{bmatrix} = \chi(a)\, \kappa(c).$$

If $\chi \neq \kappa$, then the induced representation $\mathrm{Ind}_B^G \rho$ is irreducible of dimension $q+1$. If $\chi = \kappa$, then ρ is a direct sum of a one-dimensional representation and an irreducible q-dimensional representation.

II. **Discrete Series or Cuspidal.** These representations are associated to a multiplicative character ω of $\mathbb{F}_{q^2} \cong \mathbb{F}_q(\sqrt{\delta})$, ($q$ odd), $\omega \neq \omega^q$. Piatetski-Shapiro [29] shows that the matrix entries of these representations resemble Kloosterman sums.

Now we assume that $\pi \in \hat{G}$ occurs as a subrepresentation of the left regular representation of G on $L^2(G/K)$ given by $\lambda(g)f(x) = f(g^{-1}x)$, i.e., the representation λ is the induced representation $\mathrm{Ind}_K^G 1$. Then we have a spherical function defined by

(15) $$s_\pi(x) = \frac{1}{|K|} \sum_{k \in K} \chi_\pi(kx), \quad \text{for } x \in G.$$

Here $\chi_\pi(x) = Tr(\pi(x))$ is the character of the representation π. See Diaconis [16] or Terras [42]. One also finds that the spherical function s_π is the upper left matrix entry of π for a suitable choice of basis of the underlying vector space.

Soto-Andrade [36] works out the implications of (15) quite explicitly for the projective linear group $PGL(2,\mathbb{F}_q) =$

$GL(2, \mathbb{F}_q)/Z$, where Z denotes the scalar matrices in $GL(2, \mathbb{F}_q)$; i.e., the center. For <u>principal series representations</u> $\pi = \pi_\chi$ induced from one-dimensional representations of the Borel subgroup B associated to a multiplicative character χ of $\mathbb{F}_q^*$ the corresponding spherical function is:

$$(16) \qquad s_\chi(x) = \frac{1}{|K|} \sum_{k \in K} p_\chi(kx\sqrt{\delta}), \quad x \in G,$$

where $p_\chi(z) = \chi(Imz)$, for $z \in H_q$, the finite upper half plane. This is the finite analog of Harish-Chandra's formula for the spherical functions on real non-compact symmetric spaces, where the discrete series representations do not raise their ugly/interesting heads (see Terras [41]).

The eigenvalues of the spherical function (16) are the $\lambda_1(a,b)$ defined by formula (4). We can make this more precise. If δ generates the multiplicative group $\mathbb{F}_q^*$, then the multiplicative character $\chi = \chi_b$ has the form

$$(17) \qquad \chi(\delta^u) = exp(2\pi i bu/(q-1)).$$

for $u, b \in \mathbb{Z}_{q-1}$. And $x \in G = GL(2, \mathbb{F}_q)$ must lie in some K-double coset of G having the form $S_a = S_q(\delta, a)$ as in (12).

Then, by part 3) of Theorem 1, we have the following result for the adjacency operator A_a of the graph $X_q(\delta, a)$ with $y \in S_a$:

$$(A_a \, s_{\chi_b})(x) = \sum_{k \in K} s_{\chi_b}(xky) = s_{\chi_b}(y) \, s_{\chi_b}(x).$$

This means that the <u>principal series spherical function is related to the principal series eigenvalue</u> (4) by

$$(18) \qquad \frac{1}{q+1} \lambda_1(a,b) = s_{\chi_b}(y), \quad \text{for } y \in S_a, \, a \neq 0, 4\delta.$$

The discrete series representations π_ω of $GL(2, \mathbb{F}_q)$ are associated to multiplicative characters $\omega \neq \omega^q$ of the multiplicative group $\mathbb{F}_{q^2}^* \cong \mathbb{F}_q(\sqrt{\delta})^*$. For this case Soto-Andrade [36] showed that the <u>discrete series spherical function is related</u>

to the discrete series eigenvalue (5) by

(19) $\quad \frac{1}{q+1} \lambda_s(c,\omega) = s_{\pi_\omega}(y)$, for $y \in S_a$, $a \neq 0, 4\delta$, $c = \frac{a}{\delta} - 2$.

Thanks to (18) and (19), the problem of proving the graphs $X_q(\delta, a)$ to be Ramanujan is equivalent to that of showing the spherical functions are bounded in absolute value by $\frac{2\sqrt{q}}{q+1}$, except in (18) when b=0 and always avoiding $a = 0, 4\delta$.

As we said earlier, Evans and H. Stark showed how to use a result of W. Schmidt [34] to bound the principal series eigenvalues or spherical functions. See Terras [40]. N. Katz [21] has shown how to bound the discrete series eigenvalues or spherical functions using the Grothendieck-Lefschetz trace formula on certain l-adic cohomology groups.

Next, in Figures 2-3, we show the two kinds of spherical functions on H_q using MATLAB and MATHEMATICA. Figure 2 is a surface plot of the matrix of discrete series or Soto-Andrade eigenvalues for $p=37$ produced by MATHEMATICA. The points plotted are

(20) $\quad \left[\frac{\lambda_s(c,\omega_b)}{2\sqrt{p}}, c^* \right]$, where $c^* = \begin{cases} c+2, & \text{for } -1 \leq c \leq 1, \\ c+1, & \text{for } 3 \leq c \leq p-3. \end{cases}$

Here $\omega_b(\zeta) = exp[(2\pi i b/(p^2-1))]$, for ζ a generator of the multiplicative group $\mathbb{F}_p(\sqrt{\delta})^*$ and $\lambda_s(c,\omega_b)$ is defined by (5) and (19). The index b runs from 1 to $\frac{p-1}{2}$.

Figure 3 is a surface plot of principal series or one-dimensional eigenvalues for p=37, produced using MATLAB. The points plotted are

(21) $\quad \left[\frac{\lambda_1(a,b)}{2\sqrt{p}}, a^* \right]$, for $a^* = \begin{cases} a, & a = 1, \ldots, 4\delta-1, \\ a-1, & a = 4\delta+1, \ldots, p-1 \end{cases}$ and $b = 1, \ldots, \frac{p-1}{2}$.

Here $\lambda_1(a,b)$ is defined in (4).

Figure 2. Surface Plot of Soto-Andrade or Discrete Series Eigenvalues for *p=37*. The matrix (20) is plotted. The figure was produced using MATHEMATICA.

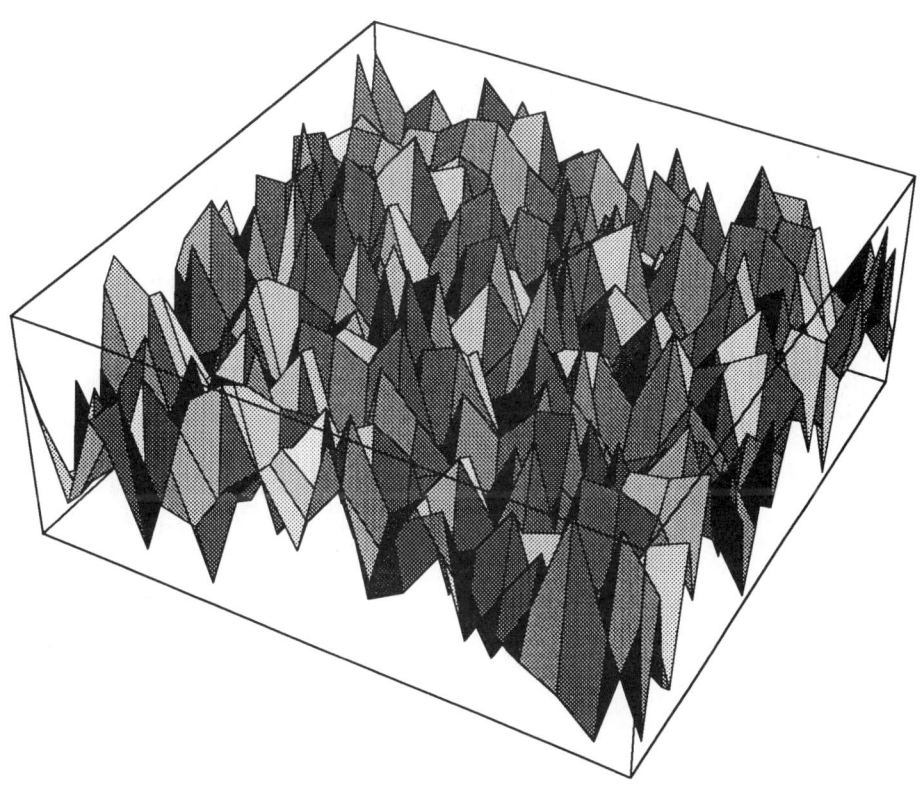

Figure 3. Surface Plot of One-Dimensional or Principal Series Eigenvalues for $p=37$. The figure was produced using MATLAB using the command mesh(m) for the matrix m defined by (21).

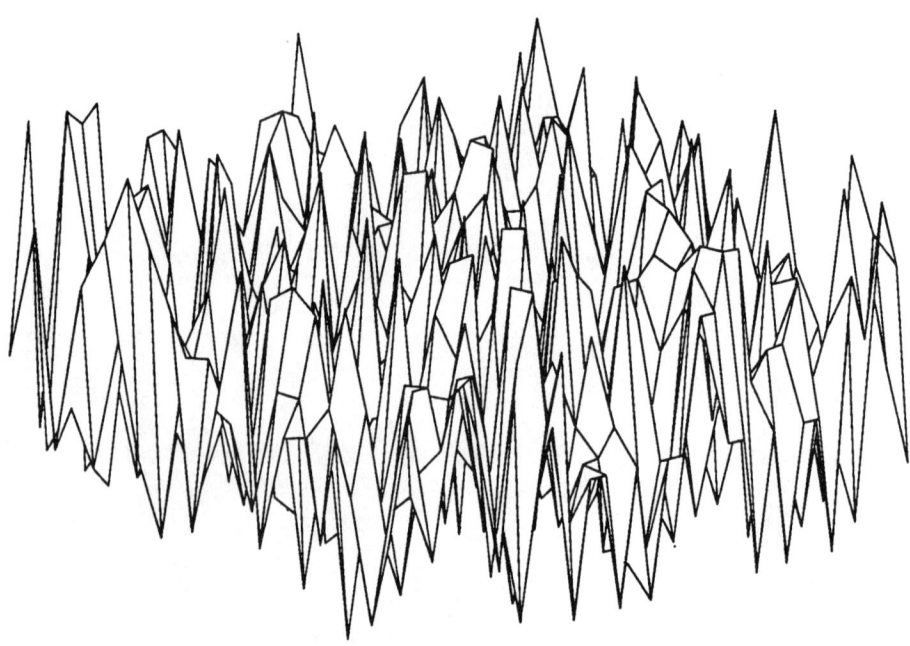

Questions.

What is the limiting distribution of the eigenvalues as q approaches infinity? Such questions were asked by Lafferty and Rockmore [22] for different graphs. We have produced similar figures to theirs using various computers. See [4]. Our figures appear to indicate that, at least if you put the eigenvalues for all $q-2$ graphs associated to G/K together, you see a continuous band $(-2\sqrt{q}, 2\sqrt{q})$. There are no gaps such as those appearing in the Lafferty and Rockmore graphs. Lafferty and Rockmore also conjecture that the behavior of principal series eigenvalues is the same as that of discrete series eigenvalues. We seem to be seeing this same phenomenon. See Figures 2 and 3 below.

One can also ask, motivated by questions in Adolphson [1] or Bollobás [7, p.351]: Do the eigenvalues approach the <u>Wigner semicircle distribution</u> (also known as the <u>Sato-Tate distribution</u>) as q approaches infinity. We obtain $\frac{q-1}{2}$ eigenvalues $\lambda \neq q+1$ of type (4) and an equal number of eigenvalues λ of type (5). When we ask whether these eigenvalues approach the semicircle distribution, we are asking whether for a set $A \subset [-2,2]$, we obtain

$$(8) \quad \frac{1}{q-1} \# \left\{ \lambda \ \bigg| \ \frac{\lambda}{\sqrt{q}} \in A \right\} \sim \frac{1}{2\pi} \int_{x \in A} \sqrt{4-x^2} \, dx, \text{ as } q \longrightarrow \infty.$$

Wigner showed that the eigenvalues of random real symmetric matrices have this distribution. So we are asking whether as q approaches infinity our adjacency operators are behaving like random symmetric matrices. Take $x = 2\cos\theta$ to obtain the Sato-Tate distribution as in Adolphson [1] and Katz [20]. See also Mehta [27] and Sarnak [31,32].

Lastly, we wonder: Can one use elementary methods to attack such problems involving distributions of exponential sums? See Katz [20] and Carlitz [11]. One could perhaps obtain weaker but still non-trivial results. Do the p-adic methods of Adolphson [1] extend to prove (7)?

Aside from our interest in these graph-theoretic questions, we also hope to make the finite symmetric spaces as concrete and well known as the real symmetric spaces. This can be useful in providing insight into the real case. See Brooks [8] and Buser [10].

Dorothy Andreoli and P. Tiu have found a clever way to attach an error-correcting code to our finite upper half plane H_p, assuming that $p = 4m+1$. The code is somewhat related to norm residue codes. They have shown that the rank of the code is at most p and that it is fixed by the projective special linear group $PSL(2,\mathbb{F}_p) = SL(2,\mathbb{F}_p)/center$. They also show that the minimum distance of the code is bounded below by $p-1$. See Tiu [44].

References

1. A. Adolphson, On the distribution of angles of Kloosterman sums, *J. für die reine und angew. Math.*, **395** (1989), 214-220.
2. J. Angel, Ph.D. Thesis, U.C.S.D., 1993.
3. J. Angel, N. Celniker, S. Poulos, A. Terras, C. Trimble, and E. Velasquez, Special functions on finite upper half planes, *Contemporary Math.*, to appear.
4. J. Angel, S. Poulos, A. Terras, C. Trimble, E. Velasquez, Spherical functions and transforms on finite upper half planes: chaotic pictures, uncertainty, traces, preprint.
5. E. Bannai and T. Ito, *Algebraic Combinatorics, I, Association Schemes*, Benjamin/Cummings, Menlo Park, CA, 1984.
6. F. Bien, Construction of telephone networks by group representations, *Notices of the Amer. Math. Soc.*, **36** (1989), 187-196.
7. B. Bollobás, *Random Graphs*, Academic, London, 1985.
8. R. Brooks, Combinatorial problems in spectral geometry, *Lecture Notes in Math.*, **1201**, Springer-Verlag, N.Y., 1986, 14-32.
9. M. Buck, Expanders and diffusers, *S.I.A.M. J. Alg. Disc. Meth.*, **7** (1986), 282-304.
10. P. Buser, Cayley graphs and planar isospectral domains, *Lecture Notes in Math.*, **1339**, Springer-Verlag, N.Y., 1988, 64-77.
11. L. Carlitz, Kloosterman sums and finite field extensions, *Acta Arith.*, **16** (1969/70), 179-193.
12. N. Celniker, Ph.D. Thesis, U.C.S.D., 1991.
13. N. Celniker, S. Poulos, A. Terras, C. Trimble and E. Velasquez, Is there life on finite upper half planes?, *Contemporary Math.*, to appear.
14. F. Chung, Diameters and eigenvalues, *J. Amer. Math. Soc.*, **2** (1989), 187-196.
15. _____, Constructing random-like graphs, *Proc. Symp, Appl. Math.*, **44** (1991), 21-55.
16. P. Diaconis, *Group Representations in Probability and Statistics*, Inst. Math. Statistics, Hayward, CA, 1988.
17. J. Friedman, Some graphs with small second eigenvalue, preprint, 1989.
18. W. Fulton and J. Harris, *Representation Theory: A First Course*, Springer-Verlag, N.Y., 1991.

19. D. Hejhal and B. Rackner, On the topography of Maass waveforms for PSL(2,$\mathbb{Z}$): experiments and heuristics, *U. of Minnesota Supercomputer Inst. Res. Rep.*, U.M.S.I., 92/162.
20. N. Katz, Sommes Exponentielle, *Astérisque,* **79** (1980), 1-209.
21. _____, Estimates for Soto-Andrade sums, preprint, 1992.
22. J. Lafferty and D. Rockmore, Fast Fourier analysis for SL_2 over a finite field and related numerical experiments, preprint.
23. A. Lubotsky, *Discrete Groups, Expanding Graphs, and Invariant Measures,* Lecture Notes, U. of Oklahoma, Norman, 1989.
24. _____, R. Phillips, and P. Sarnak, Ramanujan graphs, *Combinatorica,* **8** (1988), 261-277.
25. G. Lusztig, Symmetric spaces over a finite field, in P. Cartier et al (Eds.), *The Grothendieck Festschrift, III,* Birkhäuser, Boston, 1990, 57-81.
26. O. Martin, A.M. Odlyzko, and S. Wolfram, Algebraic properties of cellular automata, *Comm. Math. Phys.*, **93** (1984), 219-258.
27. M.L. Mehta, *Random Matrices,* Academic Press, Boston, 1991.
28. Y. Nambu, Field theory of Galois fields, in I.A. Batalin et al (Eds.), *Quantum Field Theory and Quantum Statistics, Vol. I,* Hilger, Bristol, 1987, 625-636.
29. I.I. Piatetski-Shapiro, *Complex Representations of GL(2,K) for Finite Fields K, Contemp. Math.,* **16** (1983) Amer. Math. Soc., Providence.
30. S. Poulos, Ph.D. Thesis, U.C.S.D., 1991.
31. P. Sarnak, *Some applications of modular forms,* Cambridge U. Press, Cambridge, 1990.
32. _____, Statistical properties of eigenvalues of Hecke operators, preprint.
33. P. Sarnak, Arithmetic quantum chaos, Schur Lectures, Tel Aviv, 1992.
34. W. Schmidt, *Equations over Finite Fields: An Elementary Approach, Lecture Notes in Math.,* **536**, Springer-Verlag, N.Y., 1976.
35. J.-P. Serre, Majorations de sommes exponentielles, *Astérisque,* **41-2** (1977), 111-126.
36. J. Soto-Andrade, Geometrical Gel'fand models, tensor quotients, and Weil representations, *Proc. Symp. Pure Math.,* **47**, Amer. Math. Soc., Providence, 1987, 305-316.
37. D. Stanton, *Finite Groups, Induced Representations and Orthogonal Polynomials,* U.C.S.D. Lecture Notes, La Jolla, CA, 1981.
38. H. Stark, Modular forms and related objects, *Canadian Math. Soc. Conf. Proc.,* **7** (1987), 421-455.
39. T. Sunada, Fundamental groups and Laplacians, *Lecture Notes in Math.,* **1339**, Springer-Verlag, N.Y., 1988, 248-277.

40. A. Terras, Eigenvalue problems related to finite analogues of upper half planes, in S. A. Fulling and F.J. Narcowich (Eds.), *40 More Years of Ramifications: Spectral Asymptotics and its Applications, in Discourses in Math.*,**1** (1991), Texas A&M, College Station, TX, 237-263.
41. _____, *Harmonic Analysis on Symmetric Spaces and Applications,I,II,* Springer-Verlag, N.Y., 1985, 1988.
42. _____, *Fourier Analysis on Finite Groups and Applications,* U.C.S.D. Lecture Notes, 1991-2.
43. E. Thiran, D. Verstegen and J. Weyers, p-adic dynamics, *J. Stat. Phys.,* **54** (1989), 893-913.
44. P. Tiu, Ph.D. Thesis, Dartmouth, 1992.
45. E. Velasquez, A little bit of uncertainty, preprint.
46. _____, Ph.D. Thesis, U.C.S.D., 1991.
47. A. Weil, *Collected Papers, Vols.,I-III,* Springer-Verlag, N.Y., 1979.

MATHEMATICS DEPARTMENT
University of California at San Diego,
LA JOLLA, CA 92093-0112

E-mail address: aterras@ucsd.edu